T0338540

ARACHIDONIC ACID METABOLISM AND TUMOR INITIATION

PROSTAGLANDINS, LEUKOTRIENES, AND CANCER

Series Editors: Kenneth V. Honn and Lawrence J. Marnett
Wayne State University School of Medicine
Detroit, Michigan

ARACHIDONIC ACID METABOLISM AND TUMOR INITIATION

edited by

Lawrence J. Marnett
Wayne State University
Detroit, Michigan

Martinus Nijhoff Publishing
a member of the Kluwer Academic Publishers Group
Boston/Dordrecht/Lancaster

Distributors for North America:
Kluwer Academic Publishers
190 Old Derby Street
Hingham, MA 02043

Distributors for all other countries:
Kluwer Academic Publishers Group
Distribution Centre
P.O. Box 322
3300 AH Dordrecht
The Netherlands

Library of Congress Cataloging in Publication Data

Main entry under title:

Arachidonic acid metabolism and tumor initiation.

(Prostaglandins, leukotrienes, and cancer)
Includes bibliographies and index.
1. Carcinogenesis. 2. Arachidonic acid—Metabolism.
I. Marnett, Lawrence J. II. Title: Tumor initiation.
III. Series. [DNLM: 1. Arachidonic Acids—metabolism.
2. Neoplasms—chemically induced. QZ 202 A658]
RC268.5.A68 1985   616.99'4071   85-7091
ISBN 0-89838-729-9

Printed in the United States of America

CONTENTS

CONTRIBUTORS

BERNARD B. DAVIS
Chief of Medicine
Veterans Administration
  Hospital
St. Louis, MO 63125

THOMAS E. ELING
Laboratory of Pulmonary
  Function and Toxicology
National Institute of
  Environmental Health Science
P.O. Box 12233
Research Triangle Park, NC 27709

CHARLES M. KING
Michigan Cancer Foundation
110 E. Warren Avenue
Detroit, MI 48201

ROBERT S. KRAUSS
Laboratory of Pulmonary
  Function and Toxicology
National Institute of
  Environmental Health Science
P.O. Box 12233
Research Triangle Park, NC 27709

ROGER LARSSON
Department of Forensic
  Medicine
Karolinska Institute
Box 60400
S-104 01 Stockholm
SWEDEN

LAWRENCE J. MARNETT
Department of Chemistry
Wayne State University
Detroit, MI 48202

PETER MOLDÉUS
Department of Forensic
  Medicine
Karolinska Institute
Box 60400
S-104 01 Stockholm
SWEDEN

JOHN R. RICE
Veterans Administration
  Hospital
St. Louis, MO 63125

DAVID ROSS
Department of Forensic
  Medicine
Karolinska Institute
Box 60400
S-104 01 Stockholm
SWEDEN

TERRY V. ZENSER
Departments of Medicine
  and Biochemistry
St. Louis University and
  Veterans Administration
  Medical Center
St. Louis, MO 63125

FOREWORD

Prostaglandins, Leukotrienes, and Cancer is a multi-volume series that will focus on an emerging area of cancer research. In 1968, R.H. Williams first reported that elevated prostaglandin levels are present in human medullary carcinoma. Since that time, the concept that arachidonic acid metabolites may be involved in cancer has expanded to include every aspect of the disease from cell transformation through metastasis.

Prostaglandins and leukotrienes are generic terms used to describe a family of bioactive lipids produced from unsaturated fatty acids (principally from arachidonic acid) via the cyclooxygenase and lipoxygenase pathways, respectively. Cyclooxygenase products consist of diverse products such as prostaglandin E_2 (PGE$_2$), prostacyclin (PGI$_2$) and thromboxane A_2 (TXA$_2$), whereas lipoxygenase products consist of hydroperoxy fatty acids and mono-, di- and tri-hydroxy acids including leukotrienes. The precursor fatty acids for the cyclooxygenase and lipoxygenase pathways are present in cellular phospholipids. This finding established an important control point in their biosynthesis—the release of substrate. This occurs in response to numerous stimuli that act at the cell surface. Dr. Bengt Samuelsson's extensive study of the metabolism of prostaglandins indicated that they are rapidly inactivated on a single pass through pulmonary circulation. Thus, they cannot act as circulating hormones and appear to be made on demand in or in the vicinity of target tissues leading to the concept that prostaglandins are local hormones or autocoids.

Altered production, qualitative and/or quantitative, of prostaglandins and leukotrienes has been implicated in the development of a number of disease states (e.g., atherosclerosis, inflammatory diseases, asthma). Evidence has been accumulating in the literature suggesting that prostaglandins and leukotrienes may stimulate or inhibit various steps in the complex etiology of cancer, i.e., steps in the progression from a transformed cell to a metastatic tumor. The initial volumes in this series will examine the roles of prostaglandins and leukotrienes in tumor initiation, tumor promotion, tumor cell growth and differentiation, tumor immunity, tumor metastasis and cancer therapy. We hope as this field of cancer research develops that this series, Prostaglandins, Leukotrienes, and Cancer, will provide a forum within the framework of current evidence for the synthesis of new hypotheses and discussion of controversial issues.

Kenneth V. Honn
Lawrence J. Marnett

PREFACE

It is well-established that chemical carcinogenesis is
a multi-stage process. A key step in the first stage,
initiation, is genetic alteration of somatic cells. In many
cases, this results from covalent modification of cellular
DNA by reactive carcinogens or their metabolites. Chapter 1
discusses the importance of adduct formation in the initiation
process and reviews the experimental approaches used to
detect it. Historically, most of the emphasis on chemical
initiators has been on environmental or industrial chemicals.
However, recent evidence indicates that naturally-occurring
chemicals, some of them products of normal mammalian metabolism,
can bind nucleic acid, induce mutation, and transform cells.
Arachidonic acid is oxygenated to a plethora of derivatives,
some of which are reactive electrophiles. Chapter 2 identifies
the best candidates for initiators derived from arachidonate
and examines the evidence that they modify nucleic acid.
It also introduces the concept of "cooxidation" - the enzymatic
oxidation of endogenous and xenobiotic compounds by hydroperoxide intermediates of arachidonate metabolism. This
theme is amplified in Chapter 3, which surveys the compounds
cooxidized and critically evaluates the evidence that they
are metabolically activated. Chapter 4 applies this analysis
to aromatic amines and nitrofurans, classes of potent bladder
carcinogens. It links the fields of renal physiology and
bladder carcinogenesis by the oxidative processes triggered
during arachidonic acid oxygenation. Finally, Chapter 5
examines in detail the role of cooxidation in the renal
toxicity and carcinogenicity of two widely used drugs -
acetaminophen and phenacetin.

All of the contributions in this volume are quite extensive.
This is by design. The field of arachidonic acid metabolism
in tumor initiation is a young one that has not been comprehensively reviewed. One of the hopes for this volume is that it

will stimulate investigators working in related fields of carcinogenesis to test for the involvement of arachidonic acid metabolism in their systems. This may be by isolation of a nucleic acid-modifying derivative of an unsaturated fatty acid or by testing for cooxidation of a previously unstudied class of carcinogens. The availability of a detailed literature background should aid in the design of meaningful experiments. Each chapter also attempts to critically evaluate the likely physiological significance of experimental results in the field. This is especially important in the case of prostaglandin H synthase-mediated carcinogen oxidation, which offers the possibility of explaining metabolic activation in tissues with low mixed-function oxidase activity.

Finally, it should be noted that although this volume concentrates on enzymatic pathways of arachidonate oxidation, it can be extended to non-enzymatic processes of unsaturated fatty acid oxidation such as lipid peroxidation. Accumulating evidence indicates that mechanistic parallels exist between enzymatic and non-enzymatic fatty acid oxidation and in fact, that prostaglandin and leukotriene biosynthesis can be considered a special case of enzyme-controlled lipid peroxidation. Thus, the studies described in this volume are prototypes for determining the role of lipid peroxidation in tumor initiation.

I am grateful to each of the authors for composing their contributions enthusiastically and sincerely. I am also grateful to John Battista for composing the index and to Mary Ann Vitucci, Janet Mullay, Susan Lyman and Ann Marie D'Antoni for invaluable assistance in composition.

Lawrence J. Marnett

ARACHIDONIC ACID METABOLISM AND TUMOR INITIATION

1

METABOLISM AND THE "INITIATION" OF TUMORS BY CHEMICALS

CHARLES M. KING

This presentation is based on the idea that genomic alterations by carcinogens are crucial to the production of tumors. The objective is to provide insight as to how investigators may wish to approach the identification and characterization of metabolic activation pathways, as well as the clarification of the role of these pathways in the induction of tumors. It is not intended to be an encyclopedic review, but rather a communication that reflects personal experience in the field of aromatic amines and the external influences of the carcinogenesis community. Hopefully, it will serve both as an introduction to the subsequent chapters of this book and as a framework for the investigation of the mechanisms by which other as yet unstudied compounds cause cancer.

1. CANCER AS AN EXPRESSION OF GENOMIC CHANGES

1.1. <u>Chemicals as etiological agents.</u> Most human cancer is believed to arise from environmental exposures as judged from epidemiological studies (1); the incidence of most cancers shows geographical variations that tend to be modified with relocation of populations. These geographical differences are rarely, if ever, attributable to racial influences. Rather, it appears that most tumor induction reflects the cummulative effects of exposure to a multiplicity of environmental agents. In all likelihood, these exposures come from such diverse areas as diet, occupation, medications, air pollution, the use of manufactured goods, personal habits as typified by smoking, and other activities yet to be recognized. While this complexity has hampered epidemiological efforts to resolve the etiology of most human cancer, unusual exposures to specific chemicals have clearly been implicated in the formation of several relatively

L.J. Marnett (ed.), ARACHIDONIC ACID METABOLISM AND TUMOR INITIATION.
Copyright © 1985. Martinus Nijhoff Publishing, Boston. All rights reserved.

uncommon tumors. Thus, cancer of the lung, urinary bladder, liver and renal pelvis have been associated with occupational and/or medicinal exposures (1,2). These observations demonstrate the chemical etiology of human cancer and validate efforts to elucidate the mechanisms involved in tumor induction by use of experimental animals.

1.2. Carcinogens as initiators. Cancer can be regarded as resulting from the production of a cell that is no longer restrained by normal growth control mechanisms. Chemicals that can effect these cellular changes to yield cancer, or malignant cells, are called carcinogens. A striking feature of the study of carcinogens has been the conclusion that most require metabolism for the expression of their latent tumorigenic properties. The specificities of carcinogen metabolism, and the patterns of tissue susceptibility that derive in part from these specificities, have permitted the development of experimental approaches to the clarification of the mechanisms of action of carcinogens (3). These studies have resulted in the concept that carcinogenesis is often characterized by an early irreversible event, referred to as initiation, and a subsequent, more prolonged multistep process that has come to be known as promotion. This chapter is concerned with the relationship of carcinogen metabolism to initiation, and how one can experimentally approach the study of this process. Subsequent chapters will explore the role of unsasturated fatty acid metabolism in initiation.

1.3. Chemical nature of initiation. Initiation, like cancer, is generally regarded as an irreversible process. The role of DNA in heredity has caused attention to be focused on DNA modifications as the probable crucial event in initiation. Although direct proof of altered DNA as the cause of malignancy is not available, the ability of transfected DNA from tumors to transform normal cells to a malignant state supports this idea (4). The following sections are based on the hypothesis that carcinogens alter the DNA of the target cell, and that this change

FIGURE 1

MOLECULAR EVENTS IN THE INDUCTION OF TUMORS BY CHEMICALS

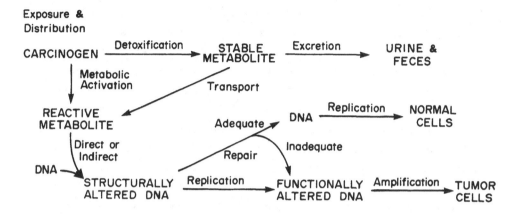

leads to the biosynthesis of DNA with abnormal base sequences. These error-containing oligonucleotide sequences, when expressed, may then produce cells with inappropriate responses to normal growth control restraints. The carcinogenic process, to this point, may be regarded as initiation. Subsequent events, which can be termed promotion, amplify these altered growth control potentials with the result that tumors result. This process is depicted diagrammatically in Figure 1.

It is apparent from Figure 1 that initiation may be a necessary, but not sufficient condition for tumor development. In reality, it is envisioned that the effectiveness of init- iation in producing tumors will depend on the relationships between the rates of DNA alteration, the rates and fidelity of the repair of these biochemical lesions and the rate of DNA synthesis. Similar considerations may apply to the process of promotion.

2. STRUCTURAL ALTERATION OF DNA

2.1. Types of alteration

2.1.1. Indirect. Genomic alterations may occur by either direct modification of the DNA, or by interference with the fidelity of DNA synthesis by indirect processes. Should the carcinogen alter the accuracy of DNA polymerases, for example, newly synthesized DNA could contain improper base sequences that were not subject to repair, and could, consequently, yield gene products that were not effective in maintaining proper control of cellular replication. Evidence that carcinogens exert their activity through such a mechanism has not been presented.

2.1.2. <u>Direct, but transitory, interactions with DNA.</u> The mutagenesis literature documents the potential for mutagenic events to arise from the non-covalent intercalation of compounds into the DNA structure in prokaryotic systems (5). This physical displacement in the absence of covalent adduct formation is sufficient to cause the misincorporation of nucleotides, thereby inducing mutations. Should this mechanism be operative in the induction of tumors, it would seem probable that unreactive agents would be capable of eliciting a carcinogenic response without undergoing metabolic transformation to products that are capable of covalent reaction with DNA. Furthermore, it would be reasonable to expect that tumors could be induced by direct application of these compounds to tissues that were known to respond to other carcinogens, as for example, the skin of mice responds to the application of polycyclic aromatic hydrocarbons (6). These characteristics are rarely, if ever, encountered. Thus, evidence for tumor induction as a consequence of non-covalent interactions with the genome is not available.

2.1.3. <u>Direct covalent interactions with DNA.</u> The two potential mechanisms for producing altered DNA that are described above are similar in that they do not postulate the existence of abnormal nucleic acid structures (e.g. adducts with xenobiotic compounds) as being responsible for initiating the carcinogenic process. Rather, the biological effects are envisioned as coming from the production of inappropriate sequences of normal bases. In contrast to this mechanism of nucleic acid alteration, for which there is no evidence of involvement in the carcinogenic process, it has been demonstrated that chemical carcinogens can react with nucleic acid to produce structures not normally produced by conventional biosynthetic pathways. While these structures could arise from the incorporation of abnormal bases into DNA, there is currently greater support from both in vitro and in vivo experiments for the idea that reactive forms of the carcinogens combine with the polynucleotide to yield covalent adducts (7,8).

2.2. Mechanisms of adduct formation

2.2.1. Reactive carcinogens. Structurally diverse compounds, as shown in Figure 2, are known to be both carcinogenic and able to react with nucleic acids under physiological conditions. Adducts recovered from tissues exposed in vivo are related to those identified in in vitro studies. Importantly, these tissues respond with the production of tumors on appropriate treatment with these carcinogens. Perhaps the best example of these relationships is seen with N-methyl-N-nitrosourea (Figure 2), a compound that is capable of reacting directly with nucleic acid and which can also induce tumors when applied directly to certain tissues such as the mammary gland (9) or urinary bladder (10).

Although the reactivity of these compounds is easily demonstrated, this reactivity complicates the demonstration of biological activity in vivo because of their instabilities. For example, it is often necessary to inject the compounds into the target tissue, instill it into a body cavity or perfuse it through the circulatory system. If administered in drinking water, on the other hand, many of these compounds will hydrolyze before they come in contact with an appropriate target cell. Criticisms are frequently leveled against studies in which compounds have been injected to induce tumors since, it is argued, direct-reacting compounds will most likely have decomposed before having established contact with crucial cellular structures if administered by routes that mimic potential human exposures. Such criticisms are not justified in the case of mechanistic, as opposed to risk-evaluation studies, but do draw attention to the low probability of environmental exposure to direct-reacting agents. However, a classic example of the effectiveness of exposure to a reactive environmental agent was

FIGURE 2

<u>Representative Carcinogens Capable of Direct Reaction</u>
<u>with Nucleic Acid under Physiological Conditions</u>

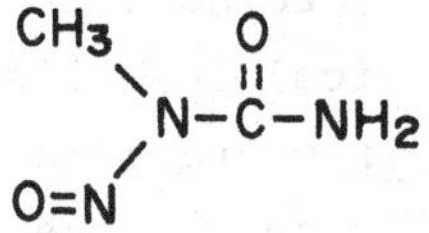

N-Methyl-N-nitrosourea

$Cl-CH_2-O-CH_2-Cl$ bis-Chloromethyl ether

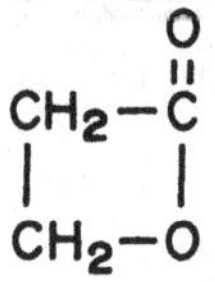

β—Propiolactone

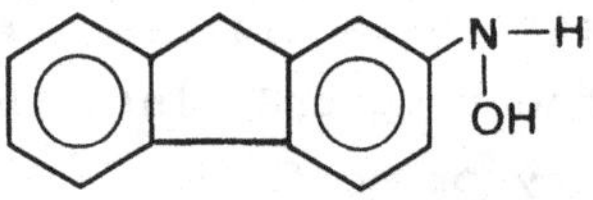

N-2-Fluorenylhydroxylamine

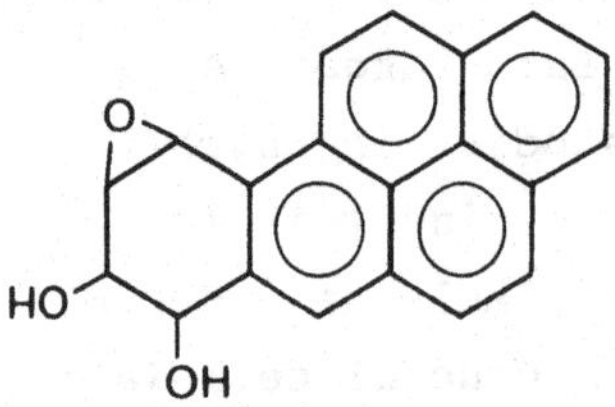

Benzo [a] pyrene-7,8 diol-9,10-epoxide

the induction of respiratory tract tumors in workers exposed occupationally to low atmospheric levels of bis-chloromethylether (11).

2.2.2. <u>Metabolic activation of carcinogens</u>. In contrast to the low probability of exposure to environmental agents that can react spontaneously with tissue macromolecules under physiological conditions, we are more likely to encounter environmental exposures to stable organic carcinogens. Almost without exception, these latter agents have been shown capable of being transformed to reactive metabolites when adequately studied (12). Some confusion has arisen concerning the nomenclature associated with this process.

Recognition that most carcinogens require metabolic conversion to derivatives that are more closely associated with the crucial molecular events of the carcinogenic process led to the introduction of terms to refer to the metabolites that were intermediates in this sequence. Thus, pro-, proximate and ultimate carcinogen have been used to indicate the relationship of a particular derivative to this metabolic process. Initially, proximate carcinogen was used to indicate a carcinogenic metabolite that was more closely related to the actual carcinogenic, i.e. reactive, species than its precursor. Confusion has arisen with this terminology because many investigators failed to recognize that proximate carcinogens were not necessarily reactive, and that reactive metabolites often designated as ultimate carcinogens could not be shown to induce tumors in experimental systems. The meaning of these terms have, therefore, become ambiguous as our knowledge of metabolic pathways has improved to the point where we have identified alternative pathways to produce the same metabolite, and additional steps and pathways have been shown to yield the actual reactive metabolites that may be involved in tumor induction. Moreover, some reactive metabolites are probably not causally involved in the carcinogenic process.

To avoid confusion and to emphasize the reactions considered most likely to be involved in the carcingenic process, it is recommended that "metabolic activation" be used to indicate that metabolic step that yields a product which is capable of reacting spontaneously with polynucleotides. This definition is based on biochemical, rather than biological criteria, in that it is based on specific reactions that can be more clearly delineated as compared to the task of establishing the relationship of the reactions to the actual production of tumors. The biochemical emphasis would minimize the questions that arise because of the production of a metabolite such as N-hydroxy-2--acetylaminofluorene that, while not able to react with nucleic acids, is able to more readily induce tumors than the parent carcinogen (i.e. 2-acetylaminofluorene) (13,14).

Previously, some investigators have taken exception to the description of a "spontaneously reactive" metabolite. This is meant to indicate that the metabolite can combine with macromolecules without further intervention of enzymes or the input of additional energy to drive the reaction. If one accepts metabolic activation as defined above, it follows that this process produces a spontaneously reactive metabolite. Adherence to such a concept would circumvent some of the difficulties encountered with the use or pro-, proximate and ultimate carcinogen.

3. CHARACTERIZATION OF METABOLIC ACTIVATION PATHWAYS

The identification of adducts and quantitation of their levels in target and non-target tissues are fundamental approaches that are required for the elucidation of the mechanisms by which chemicals initate tumor induction.

3.1. Use of radioisotopes for detection of adducts

Technical considerations (e.g. the use of radioisotopic tracers) have resulted in greater attention having been given to those pathways that yield covalent adducts, as contrasted with the production of altered nucleic acid components that do not possess structural moieties derived from the carcinogen. The earliest demonstration of protein adducts with carcinogen was made possible by the use of absorption spectroscopy to detect the presence of colored azo dyes in proteins isolated from the livers of rats treated with these compounds (15). While this

observation was pivotal in drawing attention to covalent
adducts, progress in the field was dependent primarily on the
availability of the parent carcinogens and appropriate
derivatives that were suitably labeled with radioisotopes. A
review of their use for this purpose has appeared recently (16).

The earliest adduct studies employed carbon-14 almost exclusively, because it was assumed that the label was metabolically
stable if no labeled carbon dioxide was produced in vivo.
Carbon labeling was limited, however, by the modest specific
activities that were permitted, as well as the difficulties that
were often encountered in synthesizing the desired derivatives.
More recently tritium labels have been employed, since they are
often more easily synthesized and offer the possibility of
higher specific activities. The slightly lower efficiency
usually achieved in detecting tritium by modern liquid
scintillation methods is unlikely to be justification for the
use of carbon-14.

The greater probability for the loss of tritium from organic
molecules offers both problems and opportunities. The tendency
for tritium to result in autoradiolytic degradation when incorporated at high specific activities can result in losses of
labile compounds and result in the need for repeated purification of the labeled compounds before use. On the positive side,
however, the specific loss of tritium is often useful in identifying the site of reaction, as in the loss of tritium from the
imidazole ring of purines. Elucidation of covalent adducts at
the C-8 position of guanine with arylamine moieties, for
example, was aided by establishing the loss of tritium from this
position (17). Another advantage of the use of tritium over
carbon is seen in studies with labeled acetyl derivatives in
biological systems. If carbon-14 labeled acetyl moities were to
be employed in studies with intact cells, there is the possibility that the radioisotope might be incorporated into macromolecules during their biosynthesis. Tritium-labeled acetyl
groups are less likely to result in the incorporation of label,
since metabolism of the acetate would be expected to result in
the loss of tritium to water. For example, even in the absence

of their structural identification, adducts derived in vivo from arylacetamides that had carbon-14 in the ring and tritium in the acetyl group would be expected to possess little tritium as a consequence of synthesis of the macromolecule de novo.

More recently it has been possible to employ physical and immunochemical methods for the detection of adducts (18). While these approaches offer great possibilities for the character-ization of adducts, particularly structurally specific detection at the cellular level, their use in the detection of adducts will generally require prior knowledge of structure, if not actual preparative techniques, in order to generate adducts to be used as antigen.

3.2. Selection of targets for the experimental production of adducts.

3.2.1. Macromolecules. The majority of metabolic activation experiments seek to characterize the mechanism of adduct form-ation of agents that are more lipophilic than the macromolecules of biological systems such as protein and nucleic acid. If one employs either protein or nucleic acid in order to generate adducts with reactive products of metabolic activation systems, the modified macromolecule to which the labeled agent is attached will usually exhibit extremely different solubility characteristics than the original substrate. These differences can readily be exploited to remove non-covalently bound carcinogen. For example, both protein and nucleic acid can be purified by ion-exchange chromatography or electrophoresis, procedures that can purify adducts without exposure to harsh conditions that might degrade them.

Nucleic acids, in addition to being of greater biological interest as regards their involvement in the production of chronic effects, can be subjected to solvent extraction to remove unbound organic compounds without producing irreversible changes in solubility of the macromolecule as is often the case with protein. It is possible, for example, to subject nucleic acid adducts to any purification technique that is employed for protein, as well as extracting them with such solvents as phenol

for removal of protein and organic compounds and precipitation
with ethanol. In extreme cases where one wishes to rigorously
demonstrate the absence of unbound carcinogen, the adducts can
be converted to organic salts, solubilized in organic solvents
such as methanol or toluene, subjected to liquid scintillation
for detection of radioactivity and recovered without loss of
adduct structure (14,17).

One of the justifications for the use of protein or nucleic
acid as traps of reactive products of metabolic activation, is
that they should provide a wider range of potential sites of
combination than would any one of their monomeric components.
In addition to the differences in physical characteristics,
moreover, protein and nucleic acid differ in the reactivities of
their functional groups. Thus, active derivatives that can
combine only with sulfhydryl groups will react with protein but
not with nucleic acid. As judged from one class of chemical
carcinogens, the aromatic amines, metabolites that can react
with nucleic acid can also react with protein (e.g.
N-acyloxyarylamines) (Figure 3), but aromatic amine derivatives
that can react with protein do not necessarily react with
nucleic acid (e.g. arylnitroso compounds) (17,19,20). Whether
or not this relationship holds true for other classes of
carcinogens, it would seem prudent to assume that reaction with
protein does not indicate the capability for reaction with
nucleic acid. While they are macromolecular adducts as judged
by size, the observation of radioactivity in precipitates formed
by treatment of tissue homogenates with trichloroacetic acid
would, for example, be primarily protein. To call such adducts,
macromolecular, could be misleading, since they should not be
taken as evidence for the production of nucleic acid adducts.

3.2.2. Monomeric components. The use of amino acids,
nucleosides or nucleotides can provide considerable information
regarding the nature of the functional group with which the
carcinogen has reacted (17,21,22) . Evidence of the modification of one of these monomers, even in the absence of further
structural chacterization, lends more creditable support to the
conclusion that adduct formation has taken place than is evident

FIGURE 3

METABOLIC FORMATION OF NUCLEIC ACID ADDUCTS

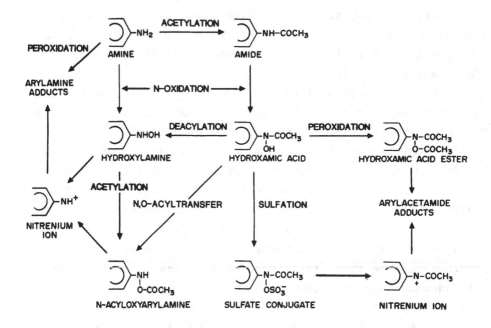

in the isolated observation of quantities of uncharacterized label in an adduct at the macromolecular level. Structural elucidation of adducts with monomers is more direct than identification of adducts with polymers, because only a limited number of functional groups are available and the monomeric adducts can be directly subjected to spectroscopic techniques without prior degradation.

There are, however, limitations to the use of monomers for detection. Unless prior information is available as to the probable functional group that is involved at the macromolecular level, it would be necessary to investigate each base for reactivity. Some investigators believe that some activated derivatives react less well with monomeric components than with macromolecules. From a qualitative perspective, at least one reactive derivative, N-acetoxy-2-acetylaminofluorene, can react with the 2-amino group and the 8-carbon of guanine of DNA, but with only the C-8 position of nucleosides or mononucleotides

(23). This difference in reactivities has been a major obstacle in the synthesis of N-2-substituted guanine derivatives.

3.3. Sources of enzymes for metabolic activation studies

3.3.1. Tissue Selection. Confronted with the problem of exploring the mechanisms of metabolic activation of a compound, the most appropriate enzyme systems are those derived from the target organ. The key characteristic of an active metabolite is that it can react with other molecules. Viewed alternatively, this reactivity implies that the compound is unstable. It follows from this relationship that the most reactive metabolites will be those that are most limited in the distance through which they will be able to exert their influence by reacting with critical cellular targets. It is important to recognize that the reactivities of activated carcinogens will vary greatly, depending on their chemical structure. Thus, some postulated activated metabolites, e.g. N-acyloxyarylamines, are sufficiently reactive that they can not be isolated by conventional synthetic approaches. These compounds must be generated in situ and studied indirectly (17,24,25). Other reactive metabolites believed to be involved in the induction of tumors, e.g. dihydrodiolexpoxide derivatives of polycyclic aromatic hydrocarbons (Figure 2), are stable enough to be synthesized and can be studied by both direct and indirect experimental approaches as described elsewhere in this book.

3.3.2. Intact cells. If it is possible to employ brief in vitro incubations of the agent with intact cells, it is often possible to demonstrate that they are capable of generating macromolecular adducts (26). Intact cells are particularly useful, since they have broader metabolic capabilities than cell-free systems and are more easily manipulated than are metabolic experiments in whole animals. Primary cells are more appropriate than those carried for prolonged periods in culture, since key metabolic capabilities may be lost during the culture process. While target cells may be more difficult to obtain, as in the cases of urinary bladder or Zymbal gland, they are more appropriate than non-target cells that may not be target cells because they lack crucial enzymes. Notwithstanding this reservation, liver cells that may not be targets have proven useful as a relatively accessible epithelial cell with broad capacities for the metabolism of xenobiotic compounds (27). Thus, liver cells can be utilized in preliminary experiments to gain insight as to the characteristics of metabolic disposition in a responsive species (28).

Metabolism in intact cells may be altered by pretreatment of the animals prior to isolation of the cells. Distinctive modifications of metabolism by hepatocytes have, as expected, been observed following treatment of the donor with such classical inducing agents as phenobarbital, Arochlor and 3-methylcholanthrene (29). Structure-activity relationships obtained from presentation of various metabolites and derivatives, alteration of endogenous cofactors such as sulfate and the addition of relatively specific inhibitors of metabolism have each been used as indirect probes of pathways that might be involved in the identification of metabolic pathways by determining the effects on adduct formation with endogenous macromolecules (26). Experiments with tissue slices have confirmed the observations obtained with isolated intact cells and also provided specific information regarding the distribution of the metabolic activation capabilities within individual cells (30).

A point of caution in the use of isolated intact cells should be noted at this point. Most investigators employing biochemical techniques are particularly careful to control the concentration of the compounds under study by adding a calculated amount of substrate to a volume of buffer. In the case of intact cells, there will be a partitioning of lipophilic agents from the aqueous media to the more lipophilic cell. The concentration within the cell, therefore, will depend on cell density rather than the calculated concentration. It is important, consequently, to ensure that comparisons are made between experiments with not only equal numbers of enzymatically active cells, but also that the number of inactive (e.g. viable and non-viable cells perhaps damaged in the isolation procedure) be controlled. Otherwise, differences in the rates and extents of metabolism may be mistakenly attributed to factors other than the actual amount of compound available to each metabolically competent cell. This phenomenon is much more likely to introduce errors in experiments with tissue slices or short-term organ cultures where it is more difficult to control the quantity and quality of tissue employed in each experiment.

A further source of information to be derived from intact cell studies is from their response to the agent. Unscheduled DNA synthesis (31) and sister chromatid exchange (32), for example, have been used as indirect assays of potential adverse effects of compounds in extrahepatic target cells. Coupled with the techniques described above, it is possible to gain information regarding the possible effectiveness of activation pathways in systems that are under metabolic controls that mimic, if not duplicate, those of in vivo exposure.

3.3.3. Cell-free enzyme preparations. Armed with information derived from knowledge of 1) the relative biological activities of a compound and its actual or potential metabolites, 2) the metabolic pathways leading to the excretion of stable low molecular products and 3) the metabolic disposition in isolated intact cell systems it is possible to attempt to devise cell--free enzyme systems that may be used for activation studies. In general, this will consist of partial purification of tissue homogenates by differential centrifugation and fortification with cofactors that will sustain the reactions that are most likely involved. This will usually be expected to involve one or more of three types of reactions (i.e. 1) oxidation or reduction, 2) conjugation or 3) hydrolysis) that would be carried out by microsomal or cytosolic components (33).

When attempting to anticipate the potential for metabolic activation of a compound, it is important to remember that one is seeking reactive compounds. In contrast to identifying stable urinary metabolites, the aim is to obtain information about derivatives that are not expected to survive such extensive transport. An alternative approach is to try to isolate the immediate precursors of a particular reactive metabolite. If these putative substrates are formed only in small quantities

and are subject to extensive metabolism, it is possible that
they will not be excreted in sufficient quantities for them to
be detected. In such cases, it is useful to administer the
precursor in order to determine whether it could be detected if
it were to be formed metabolically. Synthesis of the precursor
serves a more important role, however, in direct metabolic
activation experiments in vitro.

3.4. Experimental approach to the characterization of activation
processes. The following describes the rationale for explor-
ation of the pathways by which a compound may be activated. The
elements of experimental systems for this purpose are shown in
the left column of Table 1. The factors in the right column
point out those parameters that may be varied so as to provide
clues as to the reactions involved. The reader is directed to
the studies demonstrating the metabolic activation of aromatic
amine derivatives for examples of how this approach has been
employed in previous studies (14,17,22,24,26,34).

3.4.1. RNA adduct generation. The most convenient system for
exploring carcinogen adduct formation consists of incubations
that contain the components shown in Table 1: an enzyme preparation, a labeled substrate and a nucleic acid molecule to act as
an acceptor of the activated carcinogen. tRNA is a convenient
nucleic acid for this purpose because it is available commer-
cially, readily soluble in aqueous buffers and easily manipulated. Each of the components can be altered to provide for
structure-activity relationships, identification of the structural elements that are incorporated, etc. The addition of
cofactors, such as sulfate or acetyl donors or those required
for oxidative reactions, will result in reactions that often are
undetectable in their absence.

An important consideration in developing systems for the
demonstration of adduct formation is that the adduct should not
be exposed to extremes of either pH or temperature, since the
adduct may be unstable under these conditions. This means that
removal of the unbound carcinogen should take place without
resort to precipitation with acids (e.g. trichloroacetic acid)

Table 1. Experimental approach to the characterization of
metabolic activation systems.

System Components	Possible variable factors
Enzyme source	Intact cell
	Homogenate
	Subcellular fractions
	Particulate
	Soluble
Substrate	Parent compound
	Oxidized metabolite
	Conjugate
	Homologues
	Analogues
Cofactors	Oxidative
	Conjugative
Metabolic modifiers	Enzyme inducers
	Amino acid-specific reagents
	Enzyme-specific reagents
Biochemical targets	RNA $\pm$ nucleotide competitors
	Homopolymers
	Nucleotides/nucleosides
	DNA

that could, for example, result in depurination of DNA adducts,
or treatment with base that might hydrolyze polyribonucleotides
or open substituted imidazole rings. A convenient and effective
protocol for recovery of nucleic acids from metabolic activation
incubations involves sequential extraction with phenol and
ether, precipitation with ethanol, solution in buffer, precipitation from the aqueous solution with cetyltrimethylammonium
bromide and solution in methanol (17). The nucleic acid content

of the methanolic solution can be determined by ultraviolet
spectroscopy and the carcinogen label by liquid scintillation
spectrometry without destruction of the sample. Additional
techniques such as ion-exchange can also be included, but they
do not lend themselves to rapid assays as have been developed by
use of glass fiber filters (24).

Adduct formation in cell-free systems has one characteristic
that is not usually encountered. In most enzyme systems the
quantity of product is usually dependent on the substrate and
enzyme concentrations. In the system described above, the
amount of adduct formed will also depend on the concentration of
of nucleic acid. Thus, those nucleic acids that are poorly
soluble, e.g. DNA, will not trap reactive species as efficiently
as with higher concentrations of nucleic acid that are possible
with more soluble species such as tRNA (e.g. >10 mg/ml).

In addition to the more conventional control experiments, it
has been found useful to carry out control incubations in which
the nucleic acid is added after, rather than before, the
incubation (14,17). This technique enables one to account for
apparent adduct formation that might be due to association of
non-covalently bound metabolites or substrates with the nucleic
acid. It also will ensure that positive results of adduct
formation are not the result of contamination of the nucleic
acid with protein adducts derived from the enzyme.

3.4.2. Mononucleotide adducts. Once it has been possible to
establish a reasonable level of adduct formation by use of tRNA
as a trapping agent, it is desirable to attempt to identify the
functional group of the nucleic acid that participates in the
adduct formation. Indirect information regarding the site of
reaction can be obtained in experiments in which mononucleotides
are added to the basic incubation system (17). If the active
metabolite is capable of reacting with the mononucleotide
analogous to the nucleic acid base, that mononucleotide will
compete with the nucleic acid for the reactive species (17,35).
It may be possible, therefore, to employ systems in which the
tRNA has been replaced with the mononucleotide which has best
been able to compete with the nucleic acid in order to generate

mononucleotide adducts for characterization purposes. While nucleosides can also be used (17,24), their generally lower solubilities render them less effective traps. However, nucleoside adduct formation is less subject to interference with phosphatases that might hydrolyze either the nucleotides added as traps, or the adducts derived from them. In practice, it is best and often necessary to use both monomers.

3.4.3. Homopolymer adducts. An alternative to the use of mononucleotides is to employ homo- or copolynucleotides (36). Secondary structure may lead to adduct formation at sites on the nucleic acid that were unreactive as monomers, as noted above for esters of N-hydroxy-2-acetylaminofluorene (37). Substitution of polynucleotides with restricted base composition in activation systems would, therefore, aid in evaluating the validity of the data obtained by use of mononucleotides. In those cases where secondary structure may play a decisive role in adduct formation, it may be possible to use homopolymers to generate reference materials for adducts derived from nucleic acids.

Another value of the use of homopolymers is that similar reactive species generated by two different metabolic systems should exhibit similar specificities. In one case it has been possible to demonstrate that arylamine substitution of homopolymers by reactive species produced by prostaglandin H synthase activation of 4-aminobiphenyl are different than those derived from the activation of N-hydroxy-4-acetylaminobiphenyl by arylhydroxamic acid N,O-acyltransferase (36). Thus, adduct formation must occur via two different species with these two activation systems.

3.4.4. Adduct characterization. Experiments carried out as described in the preceeding sections should yield information as to the structural elements of the carcinogen and the nucleic acid that make up the adducts. This methodology should also permit the production of sufficient quantities of material for characterization by mass spectroscopy, nuclear magnetic resonance and ultraviolet spectroscopy. The partitioning of adducts between organic and aqueous phases as a function of pH has provided useful information on the availability of ionizable groups on the adduct (38). Combined, these techniques can provide the molecular weight of the adduct and the positions of substitution of both the carcinogen and the nucleic acid component. Applied to adducts derived from DNA they furnish the methodology to analyze adducts generated in vivo. These data also provide the basis for development of unambiguous chemical assays for the enzyme that carried out the metabolic activation and/or validates the use of possibly more convenient nucleic acid-based trapping assays.

4. SPECIFICITIES OF ACTIVATION

4.1. Chemical. The structural diversity of chemical carcinogens results in the potential generation of a wide spectrum of reactive metabolites. Several of the diverse types of reactions that might be expected to lead to the activation of carcinogens are shown in Figures 3 and 4 (39). The investigator who is attempting to identify how a compound is activated is confronted with numerous pathways that must be considered. More importantly, a single compound may have several metabolic potentials for producing adducts as is perhaps best seen in Figure 3. Historically, detection of the first pathway has tended to impede searches for or belief in alternative activation schemes. Activation of polycyclic aromatic hydrocarbons, once thought to involve a single epoxidation, was later shown to often involve a sequence of three distinct reactions (40). In the absence of data, pathways demonstrated with one carcinogen are often incorrectly assumed to be responsible for the activation of structural analogues. Consequently, the biochemical potential of a compound must be extensively evaluated in concert with attempts to determine the role of metabolic activation in tumor production.

4.2. Cellular. Deservedly, much attention has been accorded the microsomal system that is capable of oxidizing many xenobiotic compounds (e.g. 3). Some are converted to reactive species; others are processed to stable compounds that may be excreted or

FIGURE 4

POTENTIAL REACTIONS OF ACTIVATED CARCINOGENS

EPOXIDATION:

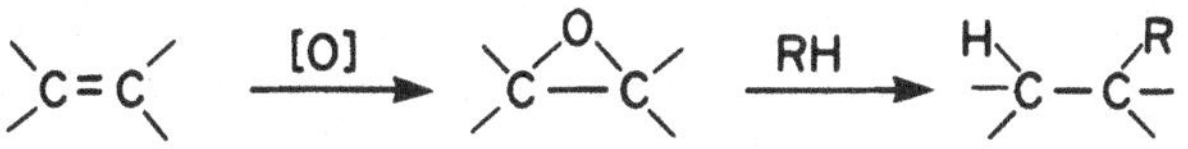

R=Nucleophilic Group

α-OXIDATION:

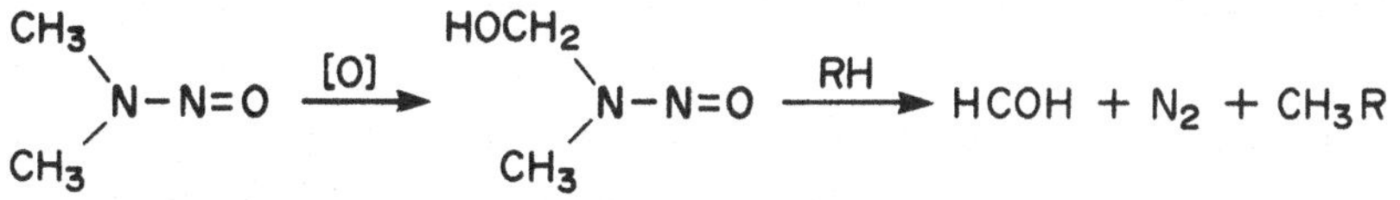

ESTERIFICATION:

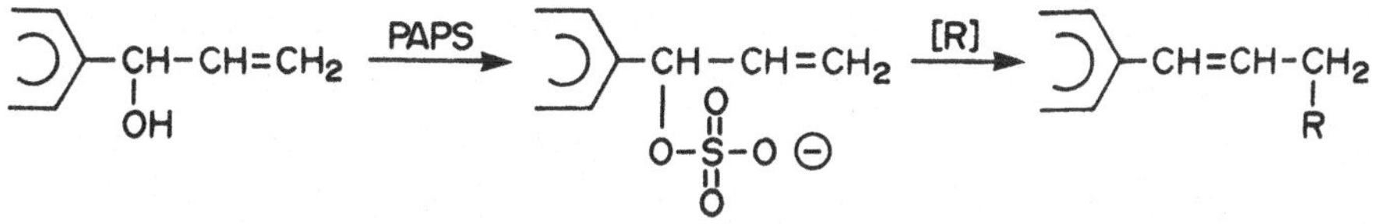

PAPS=3'-phosphoadenosine-5'-phosphosulfate

SULFHYDRYL CONJUGATION:

$$Cl-CH_2-CH_2-Cl \xrightarrow{GSH} GS-CH_2-CH_2-Cl \xrightarrow{RH} GS-CH_2-CH_2-R$$

GSH=Glutathione

activated in subsequent steps. The location of this particulate
system in the cell may determine, in part, the influence these
metabolites play in the initiation process because of the
limited distances through which many of the most active derivatives may be transported. Similar arguments may be presented
for the role of cytosolic enzymes that can conjugate certain
xenobiotics with glutathione, sulfate and acetic acid. Under
appropriate conditions these reactions can lead to the formation
of activated products (41,17,21,42).

The presence of an enzyme within a cell may not result in its
participation in the process of genome modification because it
can not compete successfully with other enzymes for substrate.
This phenomenon is probably best illustrated by the observation
that 3-methylcholanthrene can increase the levels of N-hydroxylation of 2-acetylaminofluorene in the livers of rats (43), a
metabolite with increased carcinogenicity (15). Yet, this
polycyclic aromatic hydrocarbon protects against the carcinogenicity of the parent arylacetamide while it has no effect on
the carcinogenicity of N-hydroxy-2-acetylaminofluorene (15).
Presumably other competing pathways are increased to an even
greater extent than N-hydroxylation without affecting those that
are responsible for metabolic activation of the hydroxamic acid.

Given the strides being made in the use of immunochemical and
molecular biological methodology it is likely that we will soon
have greater insight as to the intracellular location of these
enzymes, as well as their regulatory mechanisms.

4.3. <u>Tissue.</u> The carcinogenic response of a tissue depends on
exposure to a suitably reactive metabolite. As would be
expected from the different pathways of activation, the
activated metabolites differ in their reactivities. Thus,
dihydrodiolexpoxide derivatives of selected polycyclic aromatic
hydrocarbons can be synthesized, isolated and shown to induce
tumors on appropriate administration to animals (44). In
contrast, N-acyloxyarylamines can be generated <u>in situ</u>, but not
isolated (25). These marked differences in reactivity can be
expected to be reflected in more stringent requirements for

metabolic activation in the target tissues of the arylamine, since it is less likely to be able to be transported than is the diolepoxide.

Some investigators have put forward the idea that enzyme systems that are capable of metabolizing xenobiotic agents have evolved from the need of the organism to protect itself from the ingestion of deleterious materials. Clearly, metabolic systems that activate these compounds cannot be regarded as protective. On the other hand, the enzymes responsible for metabolic activation have rarely, if ever, been accorded a role in the metabolism of endogenous substrates formed by conventional biochemical pathways. The prostaglandin synthesis system considered in this book is a notable exception.

The distributions of enzymes that are capable of metabolic activation vary greatly. These unpredictable patterns of activity further support the argument that the activities that we measure are not usually involved in metabolism that is vital to the maintenance of the cells. Rather, metabolic activation is more probably the consequence of the evolutionary emergence of unusual catalytic potentials of enzymes involved in conventional biochemical pathways, as well as exposures to xenobiotic substrates. The erratic distribution these enzymes is seen within different tissues of an animal, as well as in comparisons of levels between different species. Data on the distribution of cytosolic and microsomal liver enzymes that are capable of activating N-acetyl- or N-formyl-substituted N-hydroxy-2-amino-fluorene illustrate this point, as shown in Table 2 (34). The reactive products of these activations are believed to be N-acyloxyarylamines formed by N,O-acyltransfer as shown in Figure 3. However, the cytosolic activation of N-acetylated derivatives is distinguished, in most cases, from the microsomal activation of both formylated and acetylated substrates by the resistance of the cytosolic activation to inhibition by phosphotriesters such as paraoxon. Pronounced sex-dependent differences in the tissues levels of the paraoxon-resistant enzyme have not been observed, even in a hormonally responsive tissue such as the mammary gland (45).

Table 2. Species variation of the metabolic activation of
N-acetyl- or N-formyl-substituted N-hydroxy-2-amino-
fluorene by liver cytosol or microsomes (34).

| | 2-Aminofluorene-tRNA adduct formation on incubation of tissue preparation with N-acetyl- or N-formyl-substituted substrate (nmoles 2-aminofluorene bound to tRNA/min/mg protein) | | | |
| | Cytosol | | Microsomes | |
Liver source	N-Acetyl	N-Formyl	N-Acetyl	N-Formyl
Rabbit				
RR+	0.20	0.02	0.08	0.17
Rr	0.13	0.01	0.09	0.18
rr	0.001	0.007	0.14	0.64
Hamster	0.04	0.004	0.45	0.26
Rat	0.034	0.003	0.02	0.18
Guinea pig	0.002	0.005	0.01	0.32
Mouse				
A/J	0.001	0.002	0.019	0.09
C57Bl/6J	N.D.#	0.003	0.031	0.21
Dog	N.D.	<0.001	0.003	0.05

+ Rabbits were classified according to their isoniazid or
   arylamine acetylator phenotype. RR = homozygous rapid; Rr =
   heterozygous rapid; rr = slow acetylator.
N.D. = not detectable.

A comparison of the cytosolic acyltransferase activity
between tissues within a species, and of the same tissue of
different species is shown in Table 3. Inspection shows that
the ability of a tissue to carry out a metabolic activation step
cannot be equated with its probability of developing tumors when
the animals is treated with compounds that might serve directly
or indirectly as substrates for that pathway. These data serve
to point out the potential for variability of metabolic

Table 3. Relative abilities of tissue cytosols to catalyze the metabolic activation of N-hydroxy-2-acetylaminofluorene to tRNA-bound derivatives by N,O-acyltransfer.

Tissue	Rat	Hamster	Rabbit	Guinea Pig	Monkey	Baboon	Pig	Human	Mouse, Dog
Liver	111	278	371	9	56	58	32	12	<2
Kidney	29	11	4	12					<2
Small Intestine	36	118	43	12	20			17	<2
Colon	38	31	6	10	<2			5	<2
Stomach	24	36	2	14	<2				<2
Lung	13	18	3	3	<2			2	
Mammary Gland	10								
Zymbal Gland	10								
Spleen	7	4	2	3	<2				<2
Brain	3	6	2	4					
Uterus			10						
Bladder			20						

These data, shown here as 10^{-11} mole of 2-aminofluorene bound to tRNA, were obtained by use of standardized assay conditions in which N-hydroxyl-2-acetylaminofluorene was incubated with tRNA and tissue cytosol (24,46). The values are underlined for tissues in which tumors are most commonly induced by aminofluorene derivatives.

activation, demonstrate that a direct correlation between enzyme distribution and carcinogenic susceptibility cannot be assumed, and underscore the need for assay of specific tissues of interest.

5. DOSE RESPONSE

Much concern has been expressed over the question of the relationship of dose and response in carcinogenesis. One school holds that adverse effects are likely to occur only at higher doses because adduct formation if favored only when the reactive species have exhausted the protective mechanisms of the cell. While this may be true for certain agents, studies with N,N-dimethyl-4-aminostilbene have shown that DNA adduct formation in the liver is essentially linear over an extreme dose range of several orders of magnitude (47). This elegant study clearly demonstrates that a carcinogen can alter DNA in a linear dose response over a greater range than can be effectively studied in tumor induction experiments. Experiments in vitro with intact liver cells and N-hydroxy-4-acetylamino-biphenyl showed that activation pathways for nucleic acid adduct formation could be saturated at lower concentrations than those that lead to binding with protein (26). This study demonstrated that low doses of a carcinogen may actually be more efficient at inducing nucleic acid alterations than are higher doses. Since excessive doses may be metabolized via pathways unrelated to the carcinogenic process, it may be possible to conserve expensive labelled materials by carrying out adduct studies at relatively modest dose levels.

6. CELL REPLICATION

The above two examples, when viewed from another perspective, are paradoxical, in that neither the aminostilbene or aminobiphenyl derivatives induce liver tumors under normal circumstances (48). The induction of liver and other tumors is believed to require DNA synthesis for fixation of genomic damage caused by the carcinogen as indicated in Figure 1. Since neither of these carcinogens is hepatotoxic and hepatocytes seldom undergo cell division without external stimulation, tumors are not induced by them. However, liver lesions consistent with neoplasia are induced when N-hydroxy-4-acetylamino-biphenyl is administered to rats previously subjected to partial hepatectomy and the animals are then treated with phenobarbital, i.e. cell replication is stimulated by mechanisms other than hepatotoxicity (49). Similar data have been presented for aminostilbene derivatives (50). Other related compounds, e.g. N-hydroxy-2-acetylaminofluorene, that can both alter DNA and produce a hepatotoxic response, can induce rat liver tumors in the absence of these additional factors. The hepatotoxicity induced by the fluorene derivative is believed to result from its activation via sulfate conjugation (30,51).

Tissues that undergo DNA synthesis under normal physiological conditions, as during the development of the female mammary gland, are susceptible to carcinogens that can produce nucleic acids in this tissue, without the need for tissue toxicity. The mammary gland of the male animal, though able to activate the carcinogen, may be less susceptible to mammary tumor formation in the absence of endocrine manipulation or endogenous hormonal effects of the carcinogen.

Given the link between DNA synthesis and tumorigenic response, it would seem prudent to ensure that the need for induced toxicity and/or physiological stimuli are fulfilled when evaluating the effects of new compounds. Such criteria are most likely to be met by a combination of a sufficiently high dose and identification of an appropriate target tissue.

7. ESTABLISHMENT OF MECHANISMS OF METABOLIC ACTIVATION IN CARCINOGENESIS

Successful application of the biochemical approach outlined above will present the investigator with the problem of attempting to determine whether the metabolic activation pathway is involved in tumor development. To answer this question three types of information are required. Does the target tissue have the appropriate enzymatic capability? Are the nucleic acid adducts formed in the target tissue compatible with the proposed

28

mechanism of metabolic activation? Can it be demonstrated that
the precursors of the activation step have an enhanced
carcinogenic potential as compared to derivatives less related
to the substrate involved?

7.1. Experimental design.
    7.1.1. Mode of administration. Given the goal of attempting
to determine the comparative activity of two related and
possibly metabolically interconvertible compounds, it is
desirable to administer the compounds by routes that avoid as
much metabolism by non-target tissues as possible. Systemic
administration of a mammary carcinogen in the diet, for example,
would subject the agent to possible metabolism by the flora of
the gut, the tissues of the gastrointestinal tract and the liver
prior to delivery to target tissue. Intraperitoneal injection
would circumvent the intestinal flora and tissues, but still
lead to metabolism in the liver. Direct administration to the
mammary gland, on the other hand, would increase the probability
of metabolism by the target tissue and minimize metabolism by
the other two tissues and the gut flora. This particular target
organ has been treated successfully by surgical implantation and
direct injection (45,52,53). While simpler with respect to xenobiotic metabolism, these latter methods may subject the target
tissue to mechanical trauma that may influence the carcinogenic
process. Similar direct approaches have also been employed with
the skin, trachea, colon, bladder, etc. Intravenous administration of carcinogen has proven useful in treating extrahepatic
tissues, i.e. mammary gland, with direct acting carcinogens that
would be unlikely to survive more extensive distribution and
transport (9). Thus, it is desirable and usually possible to
employ more than one method to avoid the problems of the
instability of the agent, metabolism by non-target tissues and
mechanical trauma induced by direct administration techniques.
    7.1.2. Selection of target tissue. Many carcinogens can
induce tumors in several tissues that differ in their metabolic
capacities, rate of DNA synthesis, accessibility and, consequently, susceptibility. Since the objective is to attempt to

Table 4. Metabolism and the carcinogenicity of N-hydroxy-
2-acetylaminofluorene in the Sprague-Dawley rat
(46,48,54,55)

Organ	Metabolic Activity			Tumorigenic Response
	Sulfate Conjugation	Acyltransfer	Deacylase	
Liver				
Male	+	+	+	+
Female	−	+	+	−
Mammary Gland	−	+	−	+
Sm. Intestine	−	+	+	+
Zymbal Gland	−	+	+	+

clarify the role of metabolism in the carcinogenic process, the
solution will be facilitated if the target tissue with the
simplest metabolism can be studied. The rat, for example, has a
spectrum of tissues that respond to aromatic amines, but that
differ in metabolic competence as summarized in Table 4
(46,48,54,55). These relationships permit experiments with the
mammary gland in the absence of either deacylase or sulfate
conjugation. The use of female animals for Zymbal gland (i.e.
ear duct), mammary gland or small intestine studies avoids both
sulfate conjugation and liver tumors in male animals. Liver
tumor formation might well force premature termination of an
experiment before the appearance of tumors with longer latent
periods in the other target organs, i.e. the ear duct or small
intestine, that were the object of the study. In experiments
with compounds that do not yield reactive sulfate conjugates
and, consequently, are neither hepatotoxic nor hepatocarcin-
ogenic, it is possible to employ male animals for the study of
ear duct and small intestine tumor induction without the
interference of mammary tumor development. These specific
examples point out how knowledge of the metabolic pathways and
their distribution in the animal species to be used can aid in
the design of experiments to explore the relationship of
activation to carcinogenicity.

7.2. Structural relationships to tumor induction. Under ideal
circumstances, it should be possible to synthesize and directly
administer the activated metabolite of a carcinogen to its
target organ. In practice, the inverse relationship between
stability and reactivity will usually result in the inability to
isolate the appropriate derivative and/or to adequately treat
the target tissue. Even in those cases where it is possible to
obtain reactive metabolites in relatively stable form, such as
the reactive sulfate conjugate of N-hydroxy-2-acetylamino-
fluorene (17), it has not been demonstrated that the metabolite
is capable of inducing tumors in vivo. The argument has been
made that the lack of carcinogenicity comes from the failure of
the active metabolite to be transported to the target cell.
Evidence in support of this conclusion comes from recognition
that neither the sulfate conjugate nor the N-acetoxy-2-amino-
fluorene derived from activation of N-hydroxy-2-acetylamino-
fluorene by N,O-acyltransfer is apparently sufficiently stable
to survive transport into bacterial tester strains (56,57).

A compromise solution to the dilemma posed by the technical
inability to directly implicate reactive metabolites with
tumorigenicity can sometimes be approached by the administration
of more stable precursor substrates that differ in their ability
to be activated by the enzyme system in question. A series of
arylhydroxamic acids that had different N-acyl moieties has been
synthesized and tested for their suitability as substrates for
activation by N,O-acyltransfer and for their abilities to induce
mammary tumors in Sprague-Dawley-derived female rats (58).

The study was based on the previous observations that
administration of the N-hydroxy derivative of 2-acetylamino-
fluorene, but not the amide, induced mammary tumors when
directly surgically implanted or injected into the gland
(45,53). Assay of mammary tissue preparations showed that they
possessed both paraoxon-resistant and paraoxon-sensitive
cytosolic N,O-acyltransferases (58); RNA and DNA adducts formed
in the tissue in vivo were consistent with formation by
activation of arylhydroxamic acids by N,O-acyltransfer (59,60).

The compounds selected for study carried N-formyl, N-acetyl
or N-propionyl groups on N-hydroxy-4-aminobiphenyl, or an
N-acetyl moiety on the fluorenyl analogue. Administration was
by i.p. injection. The enzyme activities were determined after
separation of the two enzymes by gel filtration. The relative
tumorigenic activities and abilities of these compounds to be
activated by the cytosolic paraoxon-sensitive and paraoxon--
insensitive acyltransferases of mammary gland are shown in
Table 5 (58).

These data clearly show that the ability of the paraoxon--
resistant acyltransferase to activate arylhydroxamic acids is
related to the abilities of the substrates to induce tumors in
the mammary gland. Surprisingly, the greater activation of the
N-formylated derivative by the paraoxon-sensitive enzyme did not
result in a related increased carcinogenic response. The
findings could be caused by the location of the paraoxon--
sensitive enzyme within the target cell, or because it is
located in a non-target cell. Another possibility is that the
greater instability of the putative formoxy derivative cannot be
transported as readily as the other derivatives. In the absence
of distribution data, it is also possible that relatively less
of the dose actually reaches the target tissue, although direct
injection of N-hydroxy-2-formylaminofluorene into the mammary
gland yielded fewer tumors than did the N-acetyl derivative
(45). Additional experiments will be required to resolve the
questions raised by this specific study, but it does highlight
another example of the need to couple biological experiments
with biochemical studies in order to establish the valid
relationships between these phenomena.

7.3. Analysis of data and potential pitfalls. Failure of tumor
induction to occur as suggested by biochemical experiments in
vitro can result from a wide range of factors, some of which
have been presented in Figure 1. In addition to the classical
factors such as metabolic activation and DNA synthesis and
repair, some tissues will be resistant to the carcinogenic
effects of a compound because of specific distribution
problems. The rabbit, for example, does not develop intestinal

Table 5.   Relationship between mammary gland activation of
           arylhydroxamic acids by N,O-acyltransfer and their
           induction of mammary tumors (58).

Compound[*]	Relative activation by N,O-acyltransfer[+]		% Tumor-bearing animals following i.p. injection
	Paraoxon-resistant	Paraoxon-sensitive	
N-OH-AAF	48	9.7	59
N-OH-AABP	6.9	<2.5	30
N-OH-PABP	5.2	<2.5	20
N-OH-FABP	<2.5	62	7
Control	--	--	3

[+] The cytosolic N,O-acyltransferases of mammary gland cytosol
were separated by gel filtration. The relative abilities of
each preparation to activate each of the four substrates is
given.

[*] N-OH-AAF = N-hydroxy-2-acetylaminobiphenyl
N-OH-AABP = N-hydroxy-4-acetylaminobiphenyl
N-OH-PABP = N-hydroxy-4-propionylaminobiphenyl
N-OH-FABP = N-hydroxy-4-formylaminobiphenyl

tumor induction on administration of aromatic amines in spite of
rather high levels of N,O-acyltransferase in this organ. This
lack of susceptibility is apparently due to the lack of excretion of aromatic amine metabolites via the bile to the gastrointestinal system (61). Thus, physiological disposition of the
carcinogen may circumvent the presence of the enzymatic capabilities of the tissue. Determination of compound distribution and
adduct formation in vivo should provide insight into apparent
anomalies such as in this case.

8. WHY DETERMINE THE MECHANISMS OF INITIATION?

Knowledge of the mechanisms by which carcinogens initiate the
carcinogenic process serves two purposes. The first is to

provide insight for the development of more rational methodologies for the evaluation of potentially harmful agents to which we may be exposed environmentally. Just as it is possible to design more sensitive carcinogenicity tests when based on an understanding of the metabolic pathways involved, this information should also permit us to develop short-term tests that reflect these mechanisms. Use of improved biochemically-based tests would also provide us with better means to apply non-- intrusive experimental approaches to human materials as a way of determining whether our studies with experimental animals is pertinent to carcinogenesis in man.

A second value to this approach is that it should aid in our understanding of the fundamental molecular mechanisms involved in malignant transformation. As we progress towards this goal, our abilities to identify and produce genomic lesions should facilitate our efforts to experimentally manipulate specific alterations of DNA that we have identified as being causally related to tumor production.

ACKNOWLEDGMENTS

This report from the A. Alfred Taubman facility was supported by National Institutes of Health grant CA 23386, awarded by the National Cancer Institute, Department of Health and Human Services; contract HEI-83-14 from the Health Effects Institute; and an institutional grant from the United Foundation of Detroit to the Michigan Cancer Foundation.

REFERENCES

1. Doll R, Peto R: The causes of cancer: quantitative estimates of avoidable risks of cancer in the United States today. J Natl Cancer Inst (66): 1191-1308, 1981.
2. Parkes HG: The epidemiology of aromatic amine cancers. In: CE Searle (ed) Chemical Carcinogens, Monograph No. 173. American Chemical Society, Washington, DC, 1976, pp 462-480.
3. King CM, Wang CY, Lee M-S, Vaught JB, Hirose M, Morton KC: Metabolic activation of aromatic amines and the induction of liver, mammary gland and urinary bladder tumors in the rat. In: Rydstrom J, Montelius J,

Bengtsson M (ed) Extrahepatic Drug Metabolism and Chemical Carcinogenesis. Elsevier Science Publishers, Amsterdam, 1983, pp 557-566.

4. Gilden RV, Rice NR: Oncogenes. Carcinogenesis (4): 791-794, 1983.

5. Sesnowitz-Horn S, Adelberg A: Proflavin treatment of E. Coli generation of frameshift mutants. Cold Spring Harbor Symp Quant Biol (33): 393-402, 1968.

6. Dipple A: Polynuclear aromatic carcinogens. In: Searle CE (ed) Chemical Carcinogens, Monograph No. 173. American Chemical Society, Washington, DC, 1976, pp 245-314.

7. Environmental Health Perspectives (49): 1-243, 1983.

8. Jerina DM, Lehr R, Schaefer-Ridder M, Yagi H, Karle JM, Thakker DR, Wood AW, Lu AYH, Ryan D, West S, Levin W, Conney AH: Bay-region epoxides of dihydrodiols: A concept explaining the mutagenic and carcinogenic activity of benzo[a]pyrene and benzo[a]anthracene. In: Hiatt HH, Watson JD, Winsten JA (eds) Origins of human cancer. Book B. Mechanisms of Carcinogenesis, Cold Spring Harbor Cenferences on Cell Proliferation, Volume 4, Cold Spring Harbor Laboratory, 1977, pp 639-658.

9. Gulino PM, Pettigrew HM, Grantham FH: N-Nitrosomethlyurea as mammary carcinogen in rats. J Natl Cancer Inst (54): 401-414, 1975.

10. Hicks RM, Wakefield JSTJ: Rapid induction of bladder cancer in rats with N-Methyl-N-nitrosourea, I. Histology. Chem-Biol Interactions (5): 139-152, 1972.

11. Nelson N: The chloro ethers: occupational carcinogens, A summary of laboratory and epidemiology studies. Ann NY Acad Sci (271): 81, 1976.

12. Miller EC: Some current perspectives on chemical carcinogenesis in humans and experimental animals: presidential address. Cancer Res (38), 1479-1496, 1978.

13. Miller EC, Miller JA, Hartman HA: N-hydroxy-2-acetyl-aminofluorene, a metabolite of 2-acetylaminofluorene with increased carcinogenic activity in the rat. Cancer Res (31): 815-824, 1961.

14. King CM, Phillips B: Enzyme-catalyzed reactions of the carcinogen N-hydroxy-2-fluorenylacetamide with nucleic acid, Science (Wash.) (159): 1351-1353, 1968.

15. Miller EC, Miller JA: The presence and significance of bound aminoazo dyes in the livers of rats fed p-dimethylaminoazobenzene. Cancer Res (7): 468-480, 1947.

16. Baird WM: The use of radioactive Carcinogens to detect DNA modifications. In: Grover PL (ed) Chemical Carcinogens and DNA, Vol. 2. Boca Raton, Florida, CRC Press, 1979, pp 59-83.

17. King CM, Phillips B: N-Hydroxy-2-fluorenylacetamide: reaction of the carcinogen with guanosine, ribonucleic acid, deoxyribonucleic acid, and protein following enzymatic deacetylation or esterification. J Biol Chem (224): 2609-2616, 1969.

18. Poirier MC, True BA, Laishes BA: Determination of 2-acetyl-aminofluorene adducts by immunoassay. Environmental Health Perspectives (49): 93-99, 1982.

19. King CM, Kriek E: The differential reactivity of the
    oxidation products of o-aminophenols towards protein and
    nucleic acid. Biochim Biophys Acta (111): 147-153, 1965.
20. King CM, Phillips B: The non-reactivity of 1,2-fluoreno-
    quinone-2-acetamide with DNA and sRNA. Biochem Pharmacol
    (17): 833-835, 1968.
21. DeBaun JR, Miller EC, Miller JA: N-Hydroxy-2-acetylamino-
    fluorene sulfotransferase: its probable role in carcino-
    genesis and protein-(methion-S-yl) binding in rat liver.
    Cancer Res (30): 577-595, 1970.
22. Bartsch H, Dworkin M, Miller J A, Miller EC: Electrophilic
    N-acetoxyaminoarenes derived from carcinogenic N-hydroxy--
    N-acetylaminoarenes by enzymatic deacetylation and
    transacetylation in liver. Biochim Biophys Acta (286):
    272-298, 1972.
23. Westra JG, Kriek E, Hittenhausen H: Identification of the
    persistently bound form of the carcinogen N-acetyl-2-amino-
    fluorene to rat liver DNA in vivo. Chem-Biol Interactions
    (15): 287-303, 1976.
24. King CM: Mechanism of reaction, tissue distribution and
    inhibition of arylhydroxamic aid acyltransferase. Cancer
    Res (34): 1503-1516, 1974.
25. Lee M-S, King CM: New synthesis of N-(guanosin-8-yl)--
    4-aminobiphenyl and its 5'monophosphate. Chem-Biol
    Interactions (34): 239-248, 1981.
26. King CM, Traub NR, Cardona RA, Howard RB: Comparative
    adduct formation of 4-aminobiphenyl and 2-aminofluorene
    derivatives with macromolecules of isolated liver
    parenchymal cells. Cancer Res (36): 2374-2381, 1976.
27. McQueen CA, Maslansky CJ, Williams GM: Role of the
    acetylation polymorphism in determining susceptibility of
    cultured rabbit hepatocytes to DNA damage by aromatic
    amines. Cancer Res (43): 3120-3123, 1983.
28. Howard PC, Casciano DA, Beland FA, Shaddock JG Jr: The
    binding of N-hydroxy-2-acetylaminofluorene to DNA and repair
    of the adducts in primary rat hepatocyte culture.
    Carcinogenesis (2): 97-102, 1981.
29. Staino N, Erickson LC, Smith CL, Marsden E, Thorgeirsson S:
    Mutagenicity and DNA damage induced by arylamines in the
    Salmonella/hepatocyte system. Carcinogenesis (4): 161-167,
    1983.
30. Shirai T, King CM: Sulfotransferase and deacetylase in
    normal and tumor-bearing liver of CD rats: autoradiographic
    studies with N-hydroxy-2-acetylaminofluorene and
    N-hydroxy-4-acetylaminobiphenyl in vitro and in vivo.
    Carcinogenesis (3): 1385-1391, 1982.
31. Wang CY, Linsmaier-Bednar EM, Garner CD, Lee M-S: Induction
    of unscheduled DNA synthesis in primary culture of dog, rat
    and mouse urothelial cells by arylamine and nitrofuran
    derivatives. Cancer Res (42): 3974-3977, 1982.
32. Wang CY, Garner CD, Lee M-S, Shirai T: O-esters of
    N-acylhydroxylamines: toxicity and enhancement of
    sister-chromatid exchange in Chinese hamster ovary cells.
    Mutation Res (88): 81-88, 1981.

33. Parke DV: The biochemistry of foreign compounds. Pergamon Press, New York, 1968, pp 34-98.
34. Glowinski IB, Savage L, Lee M-S, King CM: Relationship between nucleic acid adduct formation and deacylation of arylhydroxamic acids. Carcinogenesis (4): 67-75, 1983.
35. Irving CC, Russell LT: Synthesis of the o-glucuronide of N-2-fluorenylhydroxylamine: reaction with nucleic acids and with guanosine-5'-monophosphate. Biochemistry (9): 2471-2476, 1970.
36. Morton KC, King CM, Vaught JB, Wang CY, Lee M-S, Marnett LJ: Prostaglandin H synthase-catalyzed reaction of carcinogenic arylamines with tRNA and homopolyribonucle- otides. Biochem Biophys Res Commun (111): 96-103, 1983.
37. Kriek E, Reitsema J: Interaction of the carcinogen N-acetoxy-2-acetylaminofluorene with polyadenylic acid: dependence of reactivity on conformation. Chem-Biol Interactions (3): 397-400, 1971.
38. Moore PD, Koreeda M: Application of the change in partition coefficients with pH to the structure determination of alkyl-substituted guanosines. Biochem Biophys Res Commun (73): 459-464, 1976.
39. Grover PL (ed) Chemical Carcinogens and DNA, Volumes I and II, CRC Press, Boca Raton, Florida, 1979, pp 440.
40. Phillips DH, Sims P: Polycyclic aromatic hydrocarbon metabolites: Their reactions with nucleic acids. In: Grover PL (ed) Chemical Carcinogens and DNA, Volume II, CRC Press, Boca Raton, Florida, 1979, 29-58.
41. Rannug U, Sundvall A, Ramel C: The mutagenic effect of 1,2-dichloroethane on Salmonella typhimurium I. activation through conjugation with glutathione in vitro. Chem-Biol Interactions (20): 1-16, 1978.
42. Glowinski IB, Fysh JM, Vaught JB, Weber WW, King CM: Evidence for common genetic control of arylhydroxamic acid N-O-acyltransferase and N-acetyltransferase of rabbit liver. J Biol Chem (255): 7883-7890, 1980.
43. Lotlikar PD, Zaleski K: Ring- and N-hydroxylation of 2-acetylaminofluorene by rat liver reconstituted cytochrome P-450 enzyme system. Biochem J (150): 561-564, 1975.
44. Slaga TJ, Gleason GL, Mills G, Wald LE, Fu PP, Lee HM, Harvey RG: Comparison of the skin tumor-initiating activities of dihydrodiols and diol-epoxides of various polycyclic aromatic hydrocarbons. Cancer Res (40): 1981-1984, 1980.
45. Allaben WT, Weeks CE, Weis CC, Burger GT, King CM: Rat mammary gland carcinogenesis after local injection of N-hydroxy-N-acyl-2-aminofluorenes: relationship to metabolic activation. Carcinogenesis (3): 233-240, 1982.
46. King CM, Allaben, WT: Arylhydroxamic acid acyltransferase. In: Jakoby WB (ed) Enzymatic Basis of Detoxication. Academic Press, New York, 1980, pp 187-197.
47. Gaugler BJM, Neumann H-G: The binding of metabolites formed from aminostilbene derivatives to nucleic acids in the liver of rats. Chem-Biol Interactions (24): 355-372, 1979.

48. Clayson DB, Garner RC: Carcinogenic aromatic amines and related compounds. In: Searle CE (ed) Chemical Carcinogens, Monograph No. 173, American Chemical Society, Washington DC, 1976, pp 366-461.

49. Shirai T, Lee M-S, Wang CY, King, CM: Effects of partial hepatectomy and dietary phenobarbital on liver and mammary tumorigenesis by two N-hydroxy-N-acylaminobiphenyls in female CD rats. Cancer Res (41): 2450-2456, 1981.

50. Hilbert D, Romen W, Neumann H-G: The role of partial hepatectomy and promoters in the formation of tumors in non-target tissues of trans-4-acetylaminostilbene in rats. Carcinogenesis (4): 1519-1525, 1983.

51. Mulder GJ, Meerman JHN: Sulfation and glucuronidation as competing pathways in the metabolism of hydroxamic acids: The role of N,O-sulfonation in chemical carcinogenesis of aromatic amines. Environmental Health Perspectives (49): 27-32, 1983.

52. Dao TL: Studies on mechanism of carcinogenesis in the mammary gland. Progr Exptl Tumor Res (11): 235-261, 1969.

53. Malejka-Giganti D, Rydell RE, Gutmann HR: Mammary carcinogenesis in the rat by topical application of fluorenylhydroxamic acids and their acetates. Cancer Res (37): 111-117, 1977.

54. Irving CC: Species and tissue variations in the metabolic activation of aromatic amines. In: Griffin AC, Shaw CR (eds) Carcinogens: Identification and Mechanisms of Action, Raven Press, New York, 1979, pp 211-228.

55. Irving CC, Janss DH, Rusell LT: Lack of N-hydroxy-2-acetyl-aminofluorene sulfotransferase activity in the mammary gland and Zymbal's gland of the rat. Cancer Res (31): 387-391, 1977.

56. Wirth PJ, Thorgeirsson SS: Mechanism of N-hydroxy-2-acetyl-aminofluorene mutagenicity in the Salmonella test system. Role of N,O-acyltransferase and sulfotransferase from rat liver. Mol Pharmacol (19): 337-344, 1981.

57. Weeks CE, Allaben WT, Louie SC, Lazear EJ, King CM: Role of hydroxamic acid acyltransferase in the mutagenicity of N-hydroxy-N-2-fluorenylacetamine in Salmonella typhimurium. Cancer Res (38): 613-618, 1978.

58. Shirai T, Fysh JM, Lee M-S, Vaught JB, King CM: N-Hydroxy--N-acylarylamines: relationship of metabolic activation to biological response in the liver and mammary gland of the female CD rat. Cancer Res (41): 4346-4353, 1981.

59. King CM, Traub NR, Lortz ZM, Thissen MR: Metabolic activation of arylhydroxamic acid N-O-acyltransferase of rat mammary gland, Cancer Res (39): 3369-3372, 1979.

60. Allaben WT, Weis CC, Fullerton NF, Beland, FA: Formation and persistance of DNA adducts from the carcinogen N-hydroxy-2-acetylaminofluorene in rat mammary gland in vivo. Carcinogenesis (4): 1067-1070, 1983.

61. Irving CC, Wiseman R, Hill JT: Bilary excretion of the o-glucuronide of N-hydroxy-2-acetylaminofluorene by the rat and rabbit. Cancer Res (27): 2309-2317, 1967.

2

ARACHIDONIC ACID METABOLISM AND TUMOR INITIATION

LAWRENCE J. MARNETT

Chapter 1 of this volume describes the importance of reactions of electrophiles with nucleic acids in the initiation phase of carcinogenesis. Compounds that covalently bind to or otherwise damage DNA (e.g., by induction of strand scission) can alter the functional properties of the DNA molecule. The genetic changes that result may be a critical component of initiation. Two ways by which arachidonic acid metabolism can lead to DNA damage have been studied. In the first, metabolites of arachidonic acid bind to DNA and induce mutation or cell transformation. In the second, cooxidations by oxidizing agents generated during fatty acid hydroperoxide metabolism metabolically activate endogenous or xenobiotic compounds to derivatives that react with DNA. Both types of DNA damage occur during arachidonic acid metabolism *in vitro* and are discussed in this chapter.

1. ARACHIDONIC ACID METABOLITES AS MUTAGENS AND CARCINOGENS
1.1. Peroxides

The metabolism of arachidonic acid produces a large number of products that contain reactive functional groups (12). Some of these metabolites bind to macromolecules and induce toxic and mutagenic responses (3-5). Peroxides are intermediates in both pathways of metabolism and are reactive to protein and nucleic acid (6,7). We have tested the endoperoxide intermediates PGG_2 and PGH_2 for mutagenicity in *Salmonella typhimurium* strains TA98 and TA100 but have not detected a positive response (L. J. Marnett, unpublished results). Hydroperoxy fatty acids that are products of lipoxygenase catalysis and intermediates of leuko-

triene biosynthesis have been extensively tested in the same
strains and in TA102 and TA104, strains that detect H_2O_2 and
organic hydroperoxides as mutagens. Positive responses have not
been seen (B.N. Ames, unpublished results). The inability to
detect fatty acid hydroperoxides as mutagens may indicate that
they are not mutagenic or it may be due to a physical property of
the hydroperoxides. Perhaps the long hydrocarbon chains alter
the partitioning of the molecule to whatever site it needs to
reach to induce mutation. This is speculation, though, and at
present there is no solid evidence that any of the peroxide intermediates of arachidonate metabolism are mutagenic although they
covalently attach to macromolecules.

1.2. Malondialdehyde

The enzymatic and non-enzymatic breakdown products of arachidonate peroxides include prostaglandins, thromboxanes, prostacyclin, leukotrienes, epoxides, polyhydroxy fatty acids, enals, and
enones (1,2,5). Another product of PGH_2 metabolism is malondialdehyde (MDA) Figure 1 (8). MDA is formed by the enzymes thromboxane synthase and prostacyclin synthase and also during the
non-enzymatic oxidation of unsaturated fatty acids termed lipid
peroxidation (9-11). Its occurrence in mammalian tissue is
virtually universal. MDA was reported to be a carcinogen and

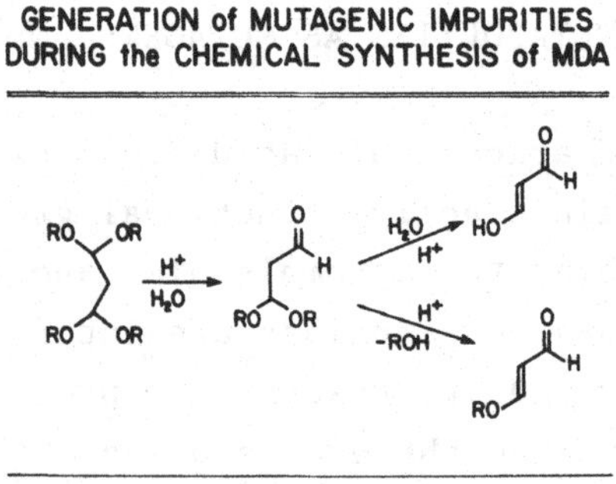

GENERATION of MUTAGENIC IMPURITIES
DURING the CHEMICAL SYNTHESIS of MDA

FIGURE 1. Origins of malondialdehyde in animal tissue.

mutagen, which suggested it might be an important mediator of
carcinogenesis (12,13). It appears, however, that most of the
mutagenicity and all of the carcinogenicity detected in the early
assays was due to side products that were formed during the preparation of MDA (Figure 2; 14,15). β-Alkoxy-acroleins, which are

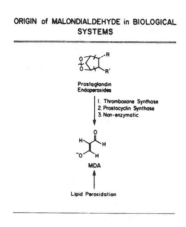

FIGURE 2. Generation of mutagenic impurities during the chemical
synthesis of MDA.

incomplete hydrolysis products of tetraalkoxypropanes, are 25-35
times more mutagenic than MDA (14). Highly purified MDA is
weakly mutagenic in *Salmonella* strain hisD3052 but it is not a
tumor initiator, promoter, or complete carcinogen when applied
topically to mouse skin (4,15). In addition, highly purified MDA
is not carcinogenic in chronic feeding studies (16). This may be
due in part to the high rate at which MDA is oxidized to CO_2
following administration to mice (17). The types of mutations
induced by MDA and its β-alkoxy-acrolein analogs (frameshift) are
quite unusual for such small molecules and they require reaction
of both carbonyl equivalents of the molecules (18).

1.3. Epoxides

Leukotriene A_4 (LTA$_4$) is an intermediate of leukotriene biosynthesis that contains an epoxide group α to a conjugated triene
functionality (eq. 1) (2). It is very unstable and adds nucleophiles at both termini of the heptatrienyl carbonium ion

eq. 1

formed by epoxide opening (2). Epoxides are usually mutagenic
in *Salmonella* strains TA98 and TA100 so one anticipates that LTA$_4$
is highly mutagenic. Preliminary experiments, however, have
failed to demonstrate any ability of LTA$_4$ to revert these strains
(L. J. Marnett and R. R. Gorman, unpublished results). One might
argue that the epoxide is so unstable (t$\frac{1}{2}$ ~ 6 sec) that it does
not reach DNA inside the bacteria when added exogenously. However, benzo[a]pyrene diolepoxides, which are unstable to
spontaneous hydrolysis (t$\frac{1}{2}$ ~ 30-120 sec), are potent mutagens in
TA98 and TA100 (19,20). The lack of mutagenicity may be due to
the physical properties of the molecule. One should recall that
prostaglandin endoperoxides and fatty acid hydroperoxides are not
mutagenic in strains that detect other organic hydroperoxides.
Electrophiles contained in molecules with fatty acid side chains
may not be capable of reaching bacterial DNA. Regardless of the
cause, there is no evidence that LTA$_4$ is mutagenic in bacteria.

1.4. Enals

Metals catalyze the conversion of fatty acid hydroperoxides
to a variety of α,β-unsaturated carbonyl compounds including
enals, enones, polyenals, and polyenones (21). 4-Hydroxy-
alkenals are major products of linoleate and arachidonate hydroperoxide breakdown and 4-hydroxy-nonenal is the principal toxic
product of lipid peroxidation (22). Acrolein and crotonaldehyde
are mutagenic in TA100 and longer chain enals are also mutagenic
in TA104 (23,24). TA100 and TA104 detect base-pair substitutions
so the types of mutations induced by enals are different than
those induced by MDA. The five carbon analog of 4-hydroxy-
nonenal, 4-hydroxy-pentenal, is mutagenic in TA104 if reduced
glutathione is added to prevent toxicity (24). Hexenal and
2,4-hexadienal are also mutagenic but nonenal, 2,4-nonadienal,

and 4-hydroxy-nonenal are not. The toxicity of the latter compounds prevents their testing at concentrations that reveal positive responses with shorter chain homologs. Also, the biological activity of the long chain enals may be reduced by the length of the hydrocarbon chains as seen with PGG_2 and LTA_4.

This brief discussion indicates that a few derivatives of unsaturated fatty acid oxygenation are mutagenic and possibly carcinogenic. A limited number of compounds have been tested but as other compounds become available in large quantities through organic synthesis a more complete picture should emerge.

2. METABOLIC ACTIVATION OF XENOBIOTICS DURING ARACHIDONIC ACID
    METABOLISM

Oxidation is intimately linked to metabolic activation of carcinogens such as polycyclic hydrocarbons (25). Polycyclic hydrocarbon oxidation in animals and man is enzyme-catalyzed and is a response to their introduction into the cellular environment (25,26). The most intensively studied enzyme of carcinogen oxidation is cytochrome P-450, which is a mixed-function oxidase that receives its electrons from NADPH via an electron transport chain (27). Some forms of this enzyme play a major role in systemic metabolism of polycyclic hydrocarbons (27). However, there are numerous examples of carcinogens that require metabolic activation but induce cancer in tissues with low mixed-function oxidase activity (28). In order to comprehensively evaluate the metabolic activation of carcinogens one must consider all cellular pathways for their oxidative activation.

Peroxidases have been implicated in carcinogenesis by *inter alia* polycyclic hydrdocarbons, aromatic amines, and estrogens (29-32). These enzymes catalyze the reduction of hydrogen peroxide and organic hydroperoxides and use a wide variety of compounds as reducing agents (eq. 2).

$$\text{eq. 2} \qquad ROOH + DH_2 \rightarrow ROH + D + H_2O$$

Although peroxidases are nearly ubiquitous, their oxidative substrate--hydrogen peroxide--is not. Its generation appears to be restricted to endocytotic vesicles of phagocytic cells, peroxisomes of oxidase-containing epithelial cells, and endoplasmic reticulum of mixed-function oxidase-containing cells that are metabolizing "uncoupling" xenobiotics (33-35). The discovery that hydroperoxides are generated as intermediates of polyunsaturated fatty acid oxygenation provides a pathway for biosynthesis of hydroperoxide substrates of peroxidases in the endoplasmic reticulum, nuclear envelope, and the soluble portion of many mammalian tissues.

The two major pathways of unsaturated fatty acid metabolism are outlined in Figure 3 for arachidonic acid. Lipoxygenase introduces one molecule of O_2 into arachidonate to form hydroperoxy eicosatetraenoic acids (HPETE's; 36). HPETE's are reduced to hydroxy eicosatetraenoic acids (HETE's) or cyclize to epoxides (37). Glutathione peroxidase and other peroxidases appear to play a role in HPETE reduction (37). The cyclooxygenase activity of PGH synthase introduces two molecules of O_2 into arachidonate to form a hydroperoxy endoperoxide (PGG_2; 6,7). PGH synthase also contains a peroxidase activity that reduces PGG_2 to a hydroxy endoperoxide (PGH_2; 38). The peroxidase appears to be a typical heme peroxidase that follows the mechanistic paradigm established for the classical peroxidase--horseradish peroxidase (39). Numerous compounds are oxidized by the peroxidase activity of PGH synthase and they are extensively reviewed in Chapter 3 of this volume and elsewhere (40).

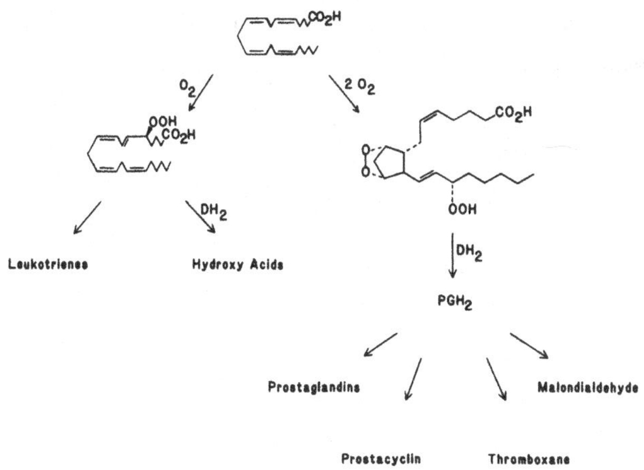

FIGURE 3. Major pathways of oxygenation of unsaturated fatty acids in animal tissue.

2.1. Benzo[a]pyrene cooxidation

We have been investigating the possibility that peroxidase-catalyzed oxidation linked to fatty acid hydroperoxide metabolism serves as a pathway for metabolic activation. One could speculate that this might be important in tissues with low mixed-function oxidase activity. We have chosen polycyclic aromatic hydrocarbons as model compounds with which to test this hypothesis (41). Incubation of BP with arachidonic acid and ram seminal vesicle microsomes, a rich source of PGH synthase, produces 1,6-, 3,6-, and 6,12-quinones as the exclusive products of oxidation (Figure 4) (42). These are the same quinones that are formed when 6-hydroxy-BP is oxidized by air or microsomes (43,44) but there is no definitive evidence that 6-hydroxy-BP is an intermediate in their formation by PGH synthase. Among all of the stable metabolites of BP, the quinones are distinctive because, unlike phenols and dihydrodiols, they are not derived from arene oxides. Thus, arene oxides do not appear to be products of BP oxidation by PGH synthase (45). Potent inhibition of PGH synthase-dependent BP oxidation by antioxidants suggests that the quinones are products of free radical reactions (42). The quinones in Figure 4 have recently been shown to be the major products of BP metabolism in human skin (46).

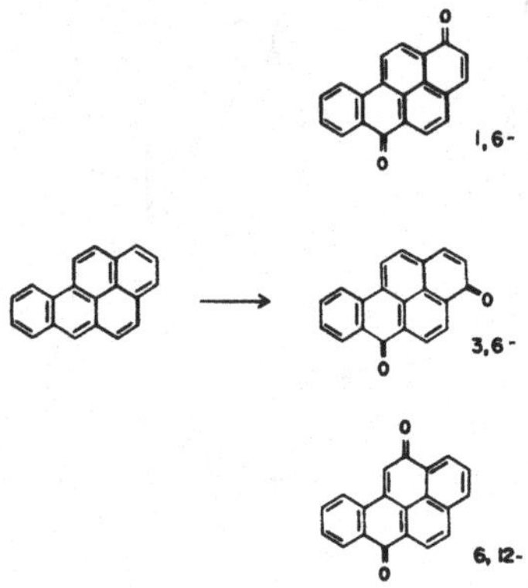

FIGURE 4. Products of BP oxidation by arachidonic acid and ram seminal vesicle microsomes.

Addition of RNA or DNA prior to oxidation of BP by PGH synthase results in substantial nucleic acid binding (45,47). Addition of RNA five minutes after initiation of oxidation leads to no covalent binding (45). This implies that the quinones do not bind to nucleic acid but rather a short-lived intermediate in their formation does. Arachidonic acid oxygenation in ram seminal vesicle microsomes is complete within two min, which suggests that the reactive intermediate is generated concurrently with PGH_2. The structures of the nucleic acid adducts have not been elucidated so the identity of the reactive intermediate is unknown.

Despite the high level of nucleic acid binding that is evident, no mutagenic species can be detected when BP is incubated with ram seminal vesicle microsomes and arachidonic acid in the presence of *Salmonella typhimurium* strains (48). The presence of *Salmonella* strains and nutrient broth in the incubations does not inhibit quinone formation. Furthermore, one of the strains employed, TA98, has been reported to detect 6-hydroxy-BP as a mutagen (49). It may be that the intermediate responsible for nucleic acid binding is too unstable to survive transit across the bacterial cell wall and membrane. Alternatively, the intermediate may bind to DNA but not induce mutation. This is unlikely because the generation of bulky adducts on a DNA molecule usually results in mutation. Although some adducts of polycyclic hydrocarbons to DNA appear to be more mutagenic than others, the differences are not greater than an order of magnitude (50,51). Thus, it seems that if adducts are formed they are mutagenic. Most of the nucleic acid adducts detected in animals following oral or topical administration of BP are derived from diolepoxides, not radical cations (52,53). This suggests that if fatty acid hydroperoxide-dependent oxidation of BP to radical cations or other free radicals plays a role in BP carcinogenesis, it is a minor one.

Two other polycyclic hydrocarbons, 3-methylcholanthrene and dimethylbenzanthracene are oxidized during arachidonate metabolism (47). Hydroxymethyl compounds that do not arise from arene oxides appear to be the products formed from 7,12-dimethylbenzanthracene (G.A. Reed, unpublished results).

2.2. 7,8-Dihydroxy-7,8-dihydrobenzo[a]pyrene (BP-7,8-diol) co-oxidation

In contrast to the results with BP, incubation of BP-7,8-diol with ram seminal vesicle microsomes and arachidonate generates a species that is strongly mutagenic to *Salmonella* strains TA98 and TA100 (Figure 5) (48). Formation of the mutagen

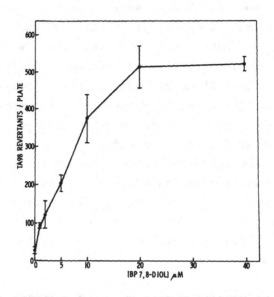

FIGURE 5. Induction of mutation in *S. typhimurium* TA98 by BP-7,8-diol, arachidonic acid, and ram seminal vesicle microsomes. Dependence on BP-7,8-diol concentration.

is inhibited by indomethacin indicating the involvement of PGH synthase. Similar experiments with BP-4,5-diol and BP-9,10-diol do not generate potent mutagens, which suggests that activation is specific for the precursor of the bay-region diolepoxide. The obvious interpretaion of these experiments is that PGH synthase catalyzes the epoxidation of BP-7,8-diol to the ultimate carcinogen BP-diolepoxide. To confirm this we identified the products of BP-7,8-diol oxidation (54,55). Theoretically, the epoxide oxygen can be introduced from either side of the molecule giving rise to syn-or anti-diolepoxides. Each epoxide hydrolyzes rapidly to a mixture of cis and trans tetrahydrotetraols (Figure 6). When incubations of BP-7,8-diol and PGH synthase are

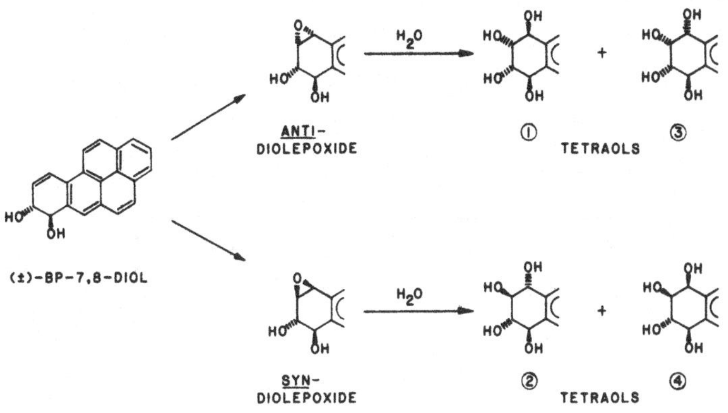

FIGURE 6. Diolepoxide products of BP-7,8-diol oxidation and their hydrolysis products.

allowed to proceed for 15 min, two products are obtained that we identified as the cis and trans tetraols derived from the anti--diolepoxide (54,55). Hydrolysis products of the syn-diolepoxide were not detected. When incubations were terminated after 3 min, a new product was detected that we identified as a methyl ether that is formed by methanolysis of the anti-diolepoxide (eq. 3) (56).

eq. 3

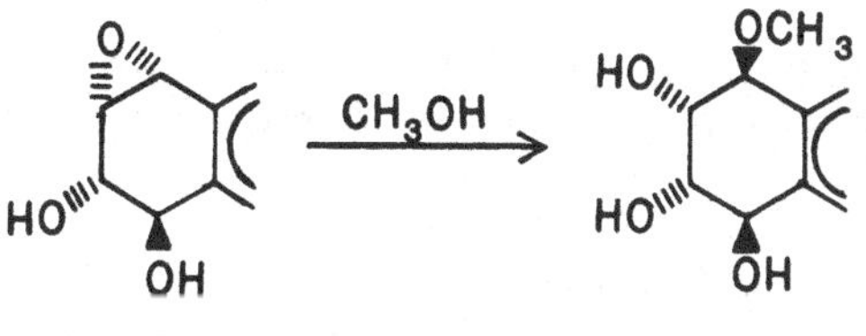

This reaction can only have occurred after termination of the reaction because there was no methanol in the incubation mixture. Additional experiments confirmed that methanolysis occurs during chromatography (reverse phase, methanol-water gradients). The detection of the methyl ether is important because it confirms that a diolepoxide is generated, survives solvent extraction, and then undergoes solvolysis on the HPLC

column. This provides direct evidence for the formation of the
anti-diolepoxide as a product of PGH synthase-dependent cooxidation of BP-7,8-diol. The correlation of the rate of BP-7,8-diol
oxidation, anti-diolepoxide formation, and mutagen generation
are shown in Figure 7 (57).

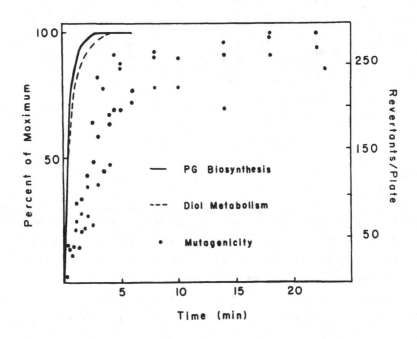

FIGURE 7. Comparison of the time courses of PGH$_2$ biosynthesis,
BP-7,8-diol metabolism, and generation of a mutagen from
BP-7,8-diol.

Further support for epoxidation of BP-7,8-diol to the anti-
diolepoxide is the identification of RNA and DNA adducts formed
as a result of incubation of BP-7,8-diol, PGH synthase, and polyguanylic acid or DNA (57,58). Following digestion of the nucleic
acid, the major guanosine and deoxyguanosine adducts were identified as arising by addition of the exocyclic amino group of guanosine to the benzylic carbon of the anti-diolepoxide (57,58).
These experiments also defined the stereochemistry of epoxidation. Both enantiomers of BP-7,8-diol are epoxidized at equal

rates to enantiomers of the _anti_-diolepoxide. The direction of
oxygen introduction is from the same side of the molecule as the
hydroxyl group at carbon-8 of BP-7,8-diol.

When the 7,8-hydroxyl groups are missing, epoxide introduction occurs from both sides of the pyrene ring. Thus, 7,8-dihydro-
benzo[a]pyrene is cooxidized by PGH synthase to a potent mutagen
that is identified by product and nucleic acid binding studies as
9,10-epoxy-7,8,9,10-tetrahydrobenzo[a]pyrene (eq. 4) (59,60). The
structures of the guanosine adducts formed in incubations containing polyguanylic acid indicate that equal amounts of epoxide
are formed by introduction of oxygen from above and below the
plane of the pyrene ring (eq. 4).

eq. 4

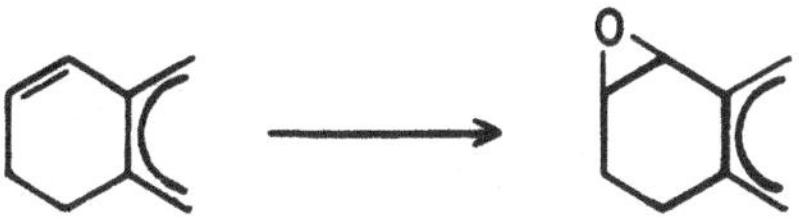

These findings indicate that PGH synthase in the presence of
arachidonate can catalyze the terminal activation step in BP carcinogenesis and that the reaction may be general for dihydrodiol
metabolites of polycyclic hydrocarbons. Guthrie _et al._ have shown
that PGH synthase catalyzes the activation of chrysene and ben-
zanthracene dihydrodiols to potent mutagens (61). As is the case
with BP, only the dihydrodiol that is a precursor to bay region
diolepoxides is activated. We have recently shown that
3,4-dihydroxy-3,4-dihydrobenzo[a]thracene is oxidized by PGH
synthase to tetrahydrotetraols derived from the _anti_-diolepoxide
(eq. 5) (T. A. Dix, V. Buck and L. J. Marnett, manuscript in
preparation).

eq. 5

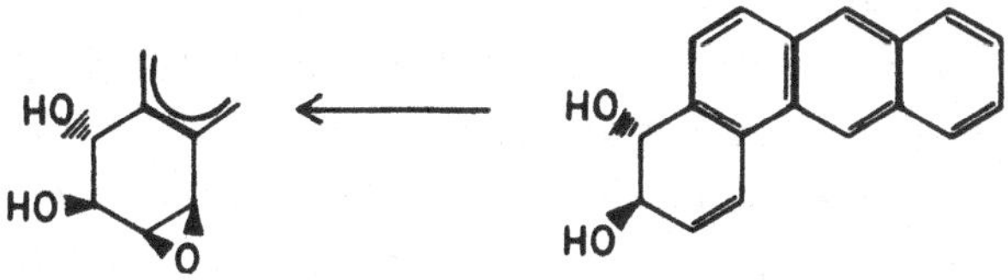

2.3. Aromatic amine cooxidation

Aromatic amines represent another important class of carcinogens that require metabolic activation. As in the case of polycyclic hydrocarbons, oxygenation is a critical step in activation, although hydroxylation rather than epoxidation is involved
(25). Chapter 1 of this volume discusses current concepts of the
activated derivatives of aromatic amines. N-hydroxylated amines
or amides with good leaving groups esterified to the hydroxyl
group serve as precursors to nitrenium ions that react with
guanine residues of nucleic acids. Pathways exist for the activation of both amines and their N-acetyl derivatives (Figure 8).
Amines are rapidly acetylated following administration to animals
and humans so activation of acetamides is of considerable
importance.

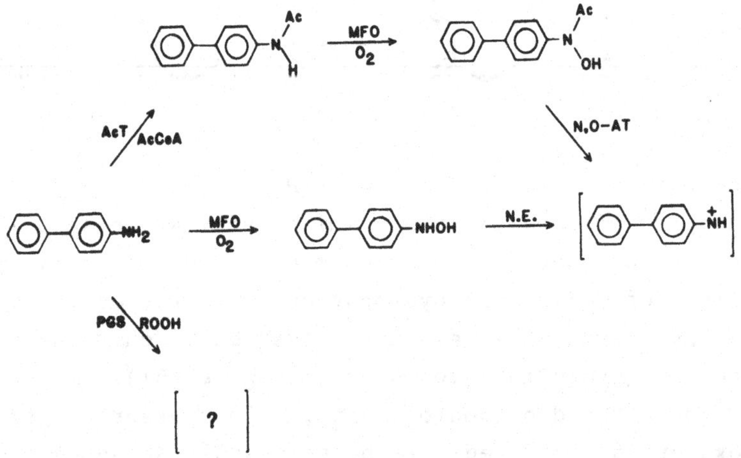

FIGURE 8. Pathways for metabolic activation of aromatic amines
by mixed-function oxidases and PGH synthase.

The extensive work of Eling, Zenser, and Moldeus *inter alia*,
reviewed elsewhere in this volume has demonstrated that aromatic
amines are excellent substrates for the peroxidase activity of
PGH synthase. Low Km's and high V_{max}'s are exhibited by the
enzyme for these substrates. N-acetyl derivatives of aromatic
amines are, by comparison, poor substrates. Therefore, PGH
synthase-catalyzed cooxidation would only seem to be potentially

important in the activation of free amines. The products of aromatic amine metabolism by peroxidases are complex and those of the PGH synthase-catalyzed reaction are no exception. Products have been identified in only a few cases that are discussed in accompanying chapters. Many studies have shown that PGH synthase induces covalent binding of aromatic amines to proteins and nucleic acids. However, the adducts have not been identified and the identities of the reactive intermediates responsible for binding are unknown.

Radical cations appear to be the primary oxidation products formed by peroxidatic metabolism of most aromatic mono- and di-amines (62). Are radical cations responsible for nucleic acid binding or are they further oxidized to nitrenium ions? We have tested for the generation of nitrenium ions during the PGH synthase-catalyzed metabolism of 4-amino-biphenyl (Figure 8) (63). Nitrenium ions derived from this compound can be generated by the N,O-acyl transferase-catalyzed rearrangement and solvo-lysis of N-hydroxy-N-acetyl-4-amino-biphenyl. When this is done in the presence of polyguanylic, polyadenylic, polycytidylic, or polyuridylic acid preferential binding to polyguanylic acid occurs. In contrast, oxidation of 4-amino-biphenyl by PGH synthase gives rise primarily to adducts to polycytidylic acid (Table 1) (63). This indicates that a nitrenium ion is not res-ponsible for nucleic acid binding detected during metabolism of 4-amino-biphenyl. Does a radical cation bind to nucleic acid or

Table 1. PGH synthase- and N,O-acyltransferase-catalyzed binding of 4-aminobiphenyl to homopolyribonucleotides.

Polynucleotide	PGH Synthase	N,O-Acyltransferase
tRNA	100	100
polyadenylic acid	<0.5	36
polyguanylic acid	6	83
polycytidylic acid	53	<0.5
polyuridylic acid	<0.5	<0.5

is it another, as yet unidentified, metabolite? Moldeus has
presented data that clearly indicate that the decrease with time
of DNA binding during PGH synthase-dependent oxidation of phenetidine corresponds closely to the decay of the radical cation of
phenetidine (Chapter 5). In this case, iminoquinones that are
formed by further oxidation of the radical cation are not
involved. In contrast, Yamazoe et al. have reported that DNA
adducts generated during PGH synthase-catalyzed oxidation of
2-napthylamine cochromatograph with adducts formed by reaction of
2-amino-1-napthol with DNA (64). They suggest that an aminophenol is generated as an intermediate in 2-napthylamine metabolism and that it is oxidized to an iminoquinone. The iminoquinone represents the reactive metabolite. Interestingly,
significant amounts of the same adducts are detected in dog
bladder following *in vivo* administration of 2-napthylamine (64).
Thus, there appear to be multiple pathways for metabolic activation of aromatic amines by peroxidases.

2.4. β-Dicarbonyl compounds

β-Dicarbonyl compounds (e.g., phenylbutazone) and other compounds with labile C-H bonds are oxidized by peroxidases (65).
The initial products are analogous to the initial products of
aromatic amine metabolism in that they are one electron oxidized
derivatives of the parent compounds (eq. 6). In contrast to

eq. 6

aromatic amines, though, these one electron oxidized forms, which
are carbon-centered free radicals, couple with O_2 to form peroxyl
radicals (eq. 7) (66). Peroxyl radicals are moderately stable

eq. 7

free radicals ($t_{\frac{1}{2}} \sim 0.1$-10 sec) that are relatively selective
oxidizing agents (67). They can abstract hydrogen to form hydroperoxides or insert oxygen into isolated double bonds. We have
detected and identified the hydroperoxide formed during peroxidatic metabolism of phenylbutazone and shown that it is
reduced to an alcohol when incubated with microsomal preparations
(P. H. Siedlik and L. J. Marnett, unpublished results). This
accounts for the overall hydroxylation of phenylbutazone that
occurs as a result of its oxidation by PGH synthase. Reed et al.
have intercepted the phenylbutazone peroxyl radical with BP-7,8-
diol which accepts the peroxyl oxygen and forms a diolepoxide
(68).

Phenylbutazone is not a carcinogen so its oxidation by PGH
synthase is of questionable relevance to carcinogenesis. However, it is a model for compounds that possess labile hydrogens
and are involved in carcinogenesis. For example, we have
recently discovered that 13-cis-retinoic acid is oxidized during
prostaglandin biosynthesis (69). The products of the reaction
and source of the oxygen suggest that a mechanism related to the
oxidation of phenylbutazone is involved in which peroxyl radicals
derived from 13-cis-retinoic acid are generated and react to form
products (69). This oxidation may be related to the chemo-
preventive activity of retinoids or to their ability to act as
tumor promoters under certain experimental conditions (70,71).

2.5. Comparison of arachidonic acid-dependent cooxidation to mixed-function oxidase-dependent cooxidation

As stated earlier, a principal driving force for the study
of peroxidatic xenobiotic oxidation is the fact that it might
represent a pathway for metabolic activation that is distinct

from the well-studied mixed-function oxidase pathways. Let us examine the evidence that these are two distinct systems and compare the properties of both.

2.5.1. Proteins involved. Mixed-function oxidases require NADPH to trigger oxidation whereas PGH synthase-dependent cooxidation is triggered by the addition of arachidonic acid (35,72). One can compare the distribution of the two enzyme systems by stimulating xenobiotic oxidation by subcellular fractions with NADPH or arachidonic acid. Such experiments show that, in general, the distribution of the two enzyme systems is complementary--tissues that have high mixed-function oxidase tend to have low PGH synthase and vice versa (40). An exception is lung which has moderate levels of both enzymes (73). Oxidation of BP-7,8-diol in this tissue is comparable when initiated by NADPH or arachidonate. Comparable levels of NADPH- and arachidonate-induced xenobiotic oxidation can also be detected in kidney homogenates. However, if the kidney is sectioned prior to homogenization, high levels of NADPH-dependent oxidation are detected in cortical preparations and very low levels in inner medullary preparations (74). Precisely opposite results are obtained for arachidonate-dependent oxidation (75). This indicates a unique complementary enzyme distribution within different regions of the same organ.

Purification of PGH synthase from bovine and ovine seminal vesicles indicates that the homogeneous protein contains both the cyclooxygenase activity that converts arachidonate to PGG_2 and the peroxidase activity that reduces PGG_2 to PGH_2 (38,76). Because the peroxidase activity is responsible for xenobiotic cooxidation, this implies that cooxidation activity is an integral part of PGH synthase. We have verified this by purifying the peroxidase from ram seminal vesicles that hydroxylates PB (39). The activity copurifies with the cyclooxygenase activity of PGH synthase and cannot be dissociated by isoelectric focusing, chromatofocusing, gel filtration, ion-exchange chromatography, or hydroxylapatite chromatography (Figure 9). Purified PGH synthase exhibits a typical b-type cytochrome visible spectrum and no maximum at 450 nm following reduction in the presence of carbon monoxide (77). Removal of heme and reconstitution regenerates a protein with cyclooxygenase and peroxidase activity that exhibits an identical visible spectrum to the holoenzyme (77).

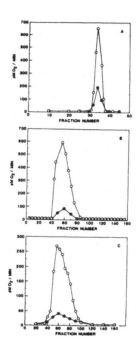

FIGURE 9. Coelution of cyclooxygenase and phenylbutazone peroxidase activities on gel filtration, chromatofocusing, and hydroxylapatite chromatography. Open circles = cyclooxygenase; closed circles = peroxidase.

Quantitation of PB peroxidase activity during chromatographic purification indicates that it is the major, if not sole peroxidase in ram seminal vesicle microsomes. In complementary experiments, we have quantitated the amounts of peroxidase activity in detergent-solubilized ram seminal vesicle microsomes by immunoprecipitation with monoclonal antibodies raised against cyclooxygenase (78). Immunoprecipitation was necessary because antibodies that inactivate the cyclooxygenase have not been prepared. The results of the experiments, summarized in Table 2, indicate that the bulk of the peroxidase activity against three different substrates immunoprecipitates with the cyclooxygenase. The fact that different amounts of activity precipitate with each substrate may be due to the fact that different assay conditions had to be employed for each one. Thus, the results of cytochemical, biochemical, and immunochemical experiments indicate that the protein responsible for PGH synthase-dependent cooxidation is distinct from mixed-function oxidases.

Table 2. Immunoprecipitation of peroxidase activities in RSVM by anticyclooxygenase

Compound	Percent activity in supernatant[a]		Percent precipitated
	Immune	Nonimmune	
Diphenyliso-benzofuran	31	72	57
Phenylbutazone	17	100	83
Epinephrine	32	62	49

[a]Percent activity remaining in supernatant after centrifugation of incubation mixtures with antibody preparations from secreting (immune) or nonsecreting (nonimmune) hybridomas.

2.5.2. Inducibility. Murota and colleagues have shown that PGH synthase is induced in cell culture by butyrate and estradiol and in liver by phenobarbital (79,80). This raises the possibility that PGH synthase is induced in a manner analogous to cytochrome P-450. Sivarajah et al. reported that treatment of enriched fractions of Clara cells and alveolar type II cells with

3-methylcholanthrene raises the level of 20:4-dependent oxidation
of BP-7,8-diol two to threefold (81). However, we have been unable to detect any increase in PGH synthase activity in the lungs
of Sprague-Dawley rats pretreated with 3-methylcholanthrene (J.
R. Battista and L. J. Marnett, unpublished results). Control
experiments demonstrate that the doses of 3-methylcholanthrene
employed were sufficient to increase lung microsomal AHH activity
three to tenfold. Another distinction between PGH synthase and
mixed-function oxidase is, therefore, their differential response
to inducing agents. The reason for the contrasting effect of
3-methylcholanthrene in cultured cells and intact animals is not
clear.

2.5.3. Substrates. PGH synthase and mixed-function oxidase
can also be differentiated on the basis of their substrates and
their oxidation products. Sivarajah et al. compared dealkyla-
tion of mono- and dialkylamines by PGH synthase and by mixed-
function oxidase. Dramatic differences in substrate specificity
were seen indicating that PGH synthase and cytochrome P-450
differ in their steric and electronic requirements for oxidation.
Similar results can be obtained in a less systematic way by comparison of the products of compounds that are oxidized by both
enzymes. Polycyclic hydrocarbons are not oxidized by PGH
synthase to arene oxides. Arene oxides are major products of
mixed-function oxidase oxidation. PB is oxidized exlusively to
4-hydroxy-PB (and the 4-hydroperoxy-PB precursor) by PGH synthase
whereas it is oxidized to oxyphenylbutazone and γ-hydroxy-phenyl-
butazone by mixed-function oxidase (Figure 10) (63,65,83). Clearcut differences in the products of oxidation of aromatic amines
and amides also exist between the two enzyme systems. In
general, PGH synthase oxidizes electron-rich molecules at the
position of highest electron density. If a molecule is easily
autoxidized it will probably be a substrate for PGH synthase and
the products will be identical to the autoxidation products. No
such simple considerations can be used to predict the products of
oxidation by cytochrome P-450. From an examination of the sub-

strate specificity of the two enzymes, it is clear that the oxidizing agent generated from PGH synthase is a weaker oxidant than the one generated from cytochrome P-450.

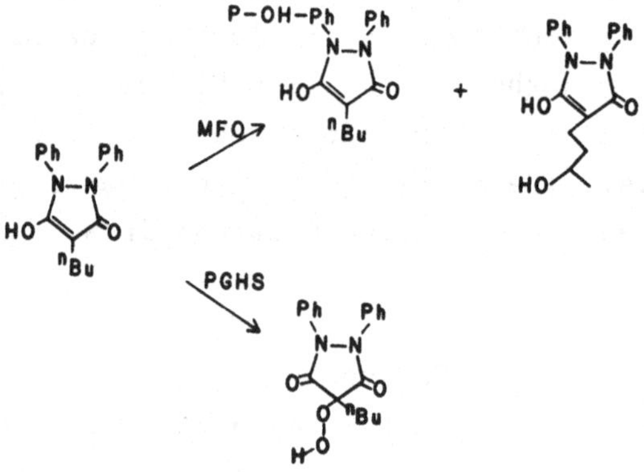

FIGURE 10. Oxidation products of phenylbutazone by mixed-function oxidase and PGH synthase.

The comparison made in this section is based on the NADPH-dependent reactions of cytochrome P-450. Cytochrome P-450 also exhibits a peroxidase activity that utilizes H_2O_2 and organic hydroperoxides (84). However, recent determinations of the turnover number of this peroxidase activity indicate that it is several orders of magnitude less efficient than "typical peroxidases" such as horseradish peroxidase toward classical peroxidase substrates (85). The same study establishes, though, that reactions such as aliphatic hydroxylation can only be performed by cytochrome P-450 and not by "typical peroxidases" (85). We have found that PGH synthase has turnover numbers comparable to those of horseradish peroxidase toward phenols and aromatic amines but it cannot carry out aliphatic hydroxylation or arene oxide formation (P. Weller and L. J. Marnett, unpublished results). The peroxidase activity of PGH synthase seems to possess the characteristics of a classical peroxidase and is quite distinct in kinetics and mechanism from the peroxidase activity of cytochrome P-450.

2.5.4. <u>Stereochemistry</u>. The final and potentially most useful distinction between PGH synthase-and mixed-function oxidase-dependent oxidation is stereochemistry. This has only been studied in detail in the case of polycyclic hydrocarbon oxygenation but it offers a potentially very powerful tool for discriminating between the two pathways of oxidation *in vitro* and, possibly *in vivo*. Polyguanylic acid was used as a chiral nucleophilic trap for diolepoxides generated from BP-7,8-diol during PGH synthase-dependent cooxidation (58). The results demonstrate that the orientation of the incoming epoxide oxygen is determined by the stereochemistry of the adjacent 8-hydroxyl group whereas stereochemistry of epoxide oxygen insertion by cytochrome P-450 is determined by the orientation of the hydrophobic aromatic side chain (86,87). The consequences of these differences are summarized in Figure 11. The (-)-enantiomer of BP-7,8-diol is oxygenated to the (+)-enantiomer of the anti-diolepoxide by both systems. This provides no basis for differentiation of the two pathways. However, the (+)-enantiomer of BP-7,8-diol is oxygenated to the (-)-enantiomer of the anti-diolepoxide by PGH synthase but to the (+)-enantiomer of the syn-diolepoxide by mixed-function oxidase. As a result, one can use the stereochemistry of epoxidation of (+)-BP-7,8-diol to identify the enzyme system that oxidizes it. We have found that this approach can be used to discriminate pathways of oxygenation even in the presence of very active mixed-function oxidase systems.

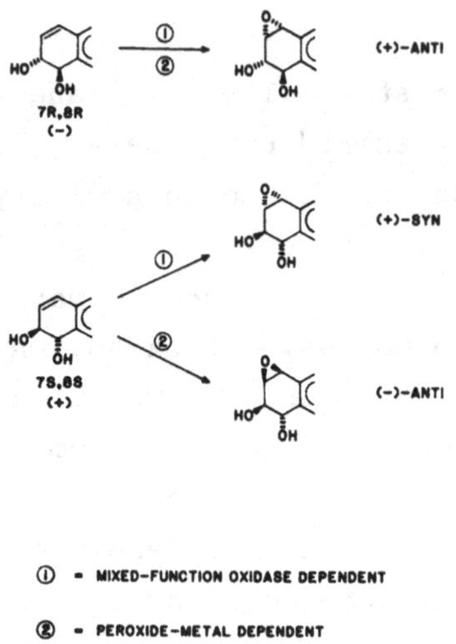

FIGURE 11. Stereochemical differences between fatty acid hydroperoxide - and mixed-function oxidase-dependent oxidation of
(±)-BP-7,8-diol.

Similar differences exist in the stereochemistry of epoxidation of 7,8-dihydrobenzo[a]pyrene (60). This compound lacks 7,8-
hydroxyl groups so the epoxide oxygen approaches from both sides
of the double bond with equal facility. However, for the reason
described above, mixed-function oxidase-dependent epoxidation is
highly stereoselective. This stereochemical difference is high-
lighted by the comparison of chromatographic profiles presented
in Figure 12.

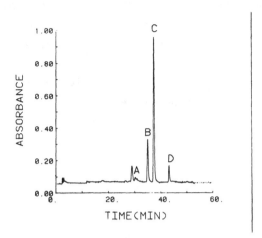

FIGURE 12. HPLC profile of guanosine adducts generated by metabolism of 7,8-dihydrobenzo[a]pyrene by PGH synthase (a) and
mixed-function oxidases (b). A and B are diastereomeric
trans-N-2 adducts and C and D are diastereomeric cis-N-2 adducts
(see 60).

2.6. Nature of oxidants generated from fatty acid hydro-
peroxides

PGH synthase contains two heme-requiring activities (38).
The cyclooxygenase component oxygenates arachidonic acid to the
hydroperoxy endoperoxide, PGG_2, and the peroxidase component
reduces PGG_2 to the hydroxy endoperoxide, PGH_2. The cyclooxygenase is inhibited by non-steroidal antiinflammatory agents
such as aspirin and indomethacin, but the peroxidase is not (87).
Both components are contained on the same 70,000 Dalton protein
(77). The presence of a peroxidase as an integral component of
PGH synthase implies that hydroperoxide-dependent oxidations are
catalyzed by this component. As a first approximation one might
expect that the mechanisms of these oxidations would be analogous
to those of other heme peroxidases.

Extensive studies have established that the catalytic cycle
for the reduction of hydroperoxides by horseradish peroxidase is
the one depicted in Figure 13 (88). The resting enzyme interacts
with the peroxide to form an enzyme-substrate complex that decom-

poses to alcohol and an iron-oxo complex that is two oxidizing
equivalents above the resting state of the enzyme. For catalytic
turnover to occur the iron-oxo complex must be reduced. The two
electrons are furnished by reducing substrates either by electron
transfer from substrate to enzyme or by oxygen transfer from
enzyme to substrate. Substrate oxidation by the iron-oxo complex
supports continuous hydroperoxide reduction. When either
reducing substrate or hydroperoxide is exhausted, the catalytic
cycle stops.

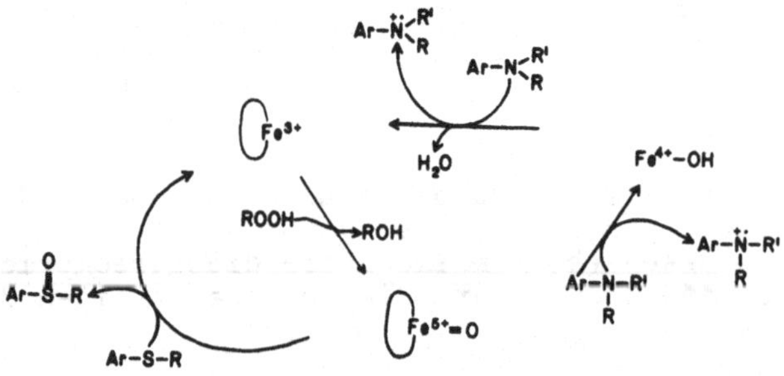

FIGURE 13. Catalytic cycle of peroxidases.

We have developed an assay to identify peroxidase reducing
substrates based on their ability to stimulate reduction of
5-phenyl-4-pentenyl-hydroperoxide (eq. 8) (89). The hydroperoxide is incubated with limiting concentrations of peroxidase

eq. 8

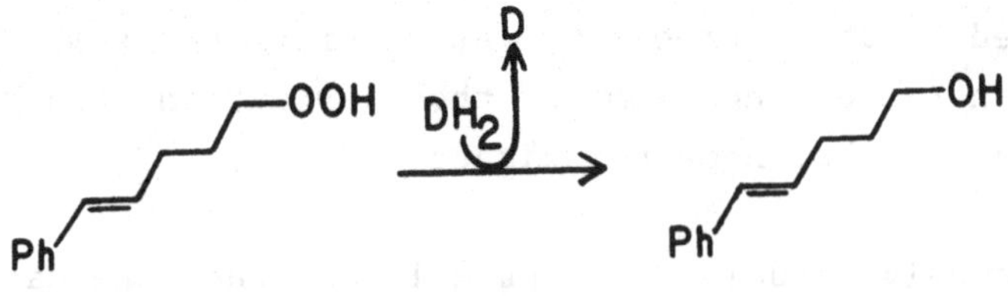

in the presence or absence of a potential reducing substrate. In the absence of reductant, catalytic reduction cannot occur and negligible quantities of alcohol are produced (the hydroperoxide and alcohol are quantitated after separation by HPLC). In the presence of a good reducing substrate catalytic turnover occurs and quantities of alcohol are produced that are stoichiometric with reducing substrate oxidized. The assay appears to be general for all plant and animal, heme and non-heme, peroxidases (89). One can rank the relative efficacy of reducing substrates using this assay. Aromatic amines, phenols, catechols, β-dicarbonyls, nitrogen heterocycles, and aromatic sulfides are good to excellent reducing substrates (89). In contrast, polycyclic hydrocarbons and dihydrodiol metabolites of polycyclic hydrocarbons are very poor to non-reducing compounds (89). Because BP and BP-7,8-diol do not stimulate hydroperoxide reduction they cannot be oxidized by higher oxidation states of the peroxidase (iron-oxo complexes). The concentrations of hydroperoxide, PGH synthase, and BP or BP-7,8-diol are analogous to those in which BP or BP-7,8-diol oxidation can be detected in ram seminal vesicle microsomes. Therefore, we conclude that the oxidizing agent that converts BP to quinones or BP-7,8-diol to diolepoxides is not an iron-oxo intermediate of peroxidase turnover.

Support for this conclusion is provided by the hydroperoxide specificity of BP oxidation (45). The scheme presented in Figure 13 requires that the same oxidizing agent is generated by reaction of H_2O_2, peroxy acids, or alkyl hydroperoxides with the peroxidase. Oxidation of any compound by the iron-oxo intermediates should be supported by any hydroperoxide that is reduced by the peroxidase. This is clearly not the case for oxidation of BP by ram seminal vesicle microsomes as the data in Figure 14 illustrate. Quinone formation is supported by fatty acid hydroperoxides but very poorly or not at all by simple alkyl hydroperoxides or H_2O_2. The fact that H_2O_2 does not support oxidation is especially significant because the same concentrations of H_2O_2 support vigorous oxidation of reducing substrates such as

aromatic amines and phenylbutazone. Therefore, we conclude that
BP and BP-7,8-diol are oxidized by a species that is not a functional intermediate of peroxidase catalysis.

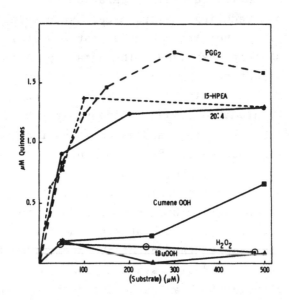

FIGURE 14. Dependence of BP oxidation by ram seminal vesicle
microsomes on the concentration of different hydroperoxides.
Abbreviations used are 20:4, arachidonic acid; 15-HPEA, 15-
hydroperoxy-eicosatetraenoic acid; t-BuOOH, t-butylhydro-
peroxide.

The oxidizing agent that is responsible for the oxygenation
of BP and BP-7,8-diol appears to be a free radical. Reaction of
fatty acid hydroperoxides with metal complexes generates alkoxyl
and peroxyl radicals that can oxidize organic molecules (90).
Incubation of fatty acid hydroperoxides with certain hemeproteins
or their prosthetic group, hematin, causes oxidation of BP to
quinones and BP-7,8-diol to diolepoxides (45,90). In the case of
BP-7,8-diol epoxidation, the source of the epoxide oxygen is
molecular oxygen; epoxidation is potently inhibited by antioxidants, and epoxidation is supported by unsaturated but not saturated fatty acid hydroperoxides (Table 3) (91). These observations are analogous to the results of microsomal incubations and
are consistent with a free radical mechanism of hydroperoxide-

dependent epoxidation. BP oxidation to quinones occurs during autoxidation of lipids initiated by enzymes or γ-irradiation (92,93).

Table 3. Dependence of hematin-catalyzed BP-7,8-diol oxidation and O_2 uptake on fatty acid hydroperoxide structure.

HYDROPEROXIDE	O_2 UPTAKE (uM)	7,8-DIOL OXIDATION VI (uM/min.)
$n-C_{16}H_{33}OOH$	0	0.16±0.02
(structure, COOH, OOH)	61±3	6.5±0.6
(structure, COOH, OOH)	160±16	12.3±0.9
(structure, COOH, HOO)	160±13	12±1.3
(structure, COOH, OOH)	160±5	12±1.6
(structure, COOCH$_3$, OOH)	0	0
(structure, COOCH$_3$, OOH)	65±1	7.0±1.1
(structure, COOCH$_3$, OOH)	240±15	16±1.8

In the case of the hematin-catalyzed reaction we have proposed that peroxyl radicals are the epoxidizing agents (90). The mechanism is illustrated in Figure 15. Hematin reduces the hydroperoxide to an alkoxyl radical that cyclizes to the adjacent double bond. The incipient carbon-centered radical couples with O_2 to form a peroxyl radical that we propose epoxidizes BP-7,8-diol. Peroxyl radicals are well-known in chemical systems to epoxidize isolated double bonds such as the 9,10-double bond of BP-7,8-diol (eq. 9) (94). However, they have been largely

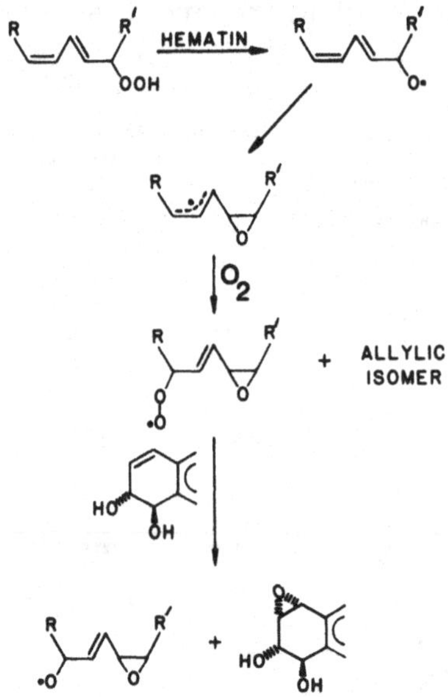

FIGURE 15. Proposed mechanism of generation of peroxyl radicals by reaction of hematin with unsaturated fatty acid hydro-peroxides.

ignored as potential oxidizing agents in biochemical systems although their half-lives (0.1-10 sec) suggest they can serve as diffusible, selective oxidants (67). The mechanism outlined in Figure 15 is consistent with all of the experimental observations and explains the requirement for a double bond in the vicinity of the hydroperoxide (Table 3). The ability of peroxyl radicals to epoxidize double bonds appears to depend upon the ability of the peroxyl radical-olefin adduct to stabilize the carbon-centered radical. Thus, 3,4-dihydroxy-3,4-dihydrobenzo[a]anthracene is oxidized to 1/6 the extent of BP-7,8-diol and aflatoxin B_1 is epoxidized to only a slight extent (95).

eq. 9 $R-O-O^{\cdot} + \left\langle\begin{smallmatrix}R'\\R''\end{smallmatrix}\right. \longrightarrow \quad R\diagup{}^O\diagdown{}^O\left\langle\begin{smallmatrix}R'\\R''\end{smallmatrix}\right. \longrightarrow \quad RO^{\cdot} + O\left\langle\begin{smallmatrix}R'\\R''\end{smallmatrix}\right.$

Peroxyl radicals are the species that propagate autoxidation of the unsaturated fatty acid residues of phospholipids (96). In addition, peroxyl radicals are intermediates in the metabolism of certain drugs such as phenylbutazone (66). Epoxidation of BP-7,8-diol has been detected during lipid peroxidation induced in rat liver microsomes by ascorbate or NADPH and during the per-oxidatic oxidation of phenylbutazone (68,97). These findings suggest that peroxyl radical-mediated epoxidation of BP-7,8-diol is general and may serve as the prototype for similar epoxida-tions of other olefins in a variety of biochemical systems. In addition, peroxyl radical-dependent epoxidation of BP-7,8-diol exhibits the same stereochemistry as the arachidonic acid-stimulated epoxidation by ram seminal vesicle microsomes. This not only provides additional evidence that the oxidizing agent in the enzymatic reaction is a peroxyl radical but also suggests that the stereochemistry of BP-7,8-diol oxidation is an important and general diagnostic probe to differentiate epoxidation by mixed-function oxidases and by peroxyl radicals.

2.7. Significance of arachidonic acid-dependent cooxidation

What is the significance of arachidonic acid-dependent xeno-biotic metabolism? Experiments described above firmly establish that it can cause metabolic activation *in vitro*. Dihydrodiol metabolites of polycyclic hydrocarbons are oxidized to dihydro-diol epoxides that represent the ultimate carcinogenic forms of the parent hydrocarbons. Interestingly, only dihydrodiols that form bay-region diolepoxides are activated by PGH synthase; no activation of other polycyclic hydrocarbon dihydrodiols occurs (48,61). Arachidonate-dependent cooxidation is essentially an activation pathway specific for generation of bay-region diolepoxides. Work described in Chapter 3 indicates that aromatic amines can also be oxidized to mutagenic derivatives although the identity of the mutagenic derivatives is at present uncertain.

Multiple pathways for arachidonate-dependent oxidation exist. In some cases, the oxidizing agent is a higher oxidation state of the peroxidase component of PGH synthase. This is true for molecules that are easily oxidized such as aromatic amines or sulfides. To a degree one can predict whether a molecule will be oxidized by the peroxidase from its one electron oxidation potential. Molecules that undergo autoxidation readily are also usually oxidized by the peroxidase. The products of PGH synthase-catalyzed oxidation are normally identical to the autoxidation products.

Compounds that are not oxidized by the peroxidase component of PGH synthase can be oxidized by free radicals--peroxyl or possibly alkoxyl radicals--that are generated by metal-catalyzed reduction or oxidation of the hydroperoxide intermediates of arachidonic acid oxygenation. This is a pathway of oxidation that is unique to enzymes that generate fatty acid hydroperoxides. The reactivity of most xenobiotics toward peroxyl and alkoxyl radicals, as well as the reaction products, are unknown, although some model chemistry is available. Thus it is more difficult to predict whether a molecule will be oxidized by the free radical-dependent pathway. Regardless of whether the oxidizing agent is a peroxidase higher oxidation state or a free radical, the products of oxidation are, in many cases, distinct from those generated by the action of mixed-function oxidases. Thus arachidonic acid-dependent cooxidation constitutes a pathway for xenobiotic metabolism and metabolic activation that is complementary to classical pathways.

PGH synthase occupies a position at the interface of peroxidase chemistry and free radical chemistry and can trigger metabolic activation by both mechanisms. The peroxidase pathway activates compounds such as diethylstilbestrol and aromatic amines whereas the free radical pathway activates polycyclic hydrocarbons. Both pathways require synthesis of hydroperoxide in order to trigger oxidation. The rate-limiting step in hydroperoxide synthesis is release of arachidonic acid from phospholipid storage (98,99). Release is catalyzed by phospholipases and is stimulated by agents that act at the cell surface such as hormones, ionophores, tumor promoters, etc. (100). Arachidonic acid-dependent cooxidation is, therefore, a pathway that links events at the cell surface to intracellular oxidation of xenobiotics. It is also a model for oxidation of xenobiotics by other peroxidases and by free radicals. There are few reports of xenobiotic metabolism by peroxyl or alkoxyl free radicals but the potential is enormous. Unsaturated fatty acids are present in all cells to some extent and, in fact, are quite abundant in most cells. For example, ventricular myocardial muscle contains 14.1 μmol linoleic and arachidonic acids per gram wet weight tissue (101). Both fatty acids are quite susceptible to lipid peroxidation which generates peroxyl radicals capable of oxidizing certain xenobiotics, e.g., epoxidizing BP-7,8-diol. Inhibition of lipid peroxidation is obviously a task that must be constantly performed by cells to prevent tissue destruction and xenobiotic metabolism. The turnover of only 0.1% of the unsaturated fatty acid residues of cells could generate a very significant amount of peroxyl radicals inside membrane regions of cells where many xenobiotics are dissolved.

Metabolism of aromatic amines and polycyclic hydrocarbon metabolites has been detected during arachidonate oxygenation in intact cells and in cultured trachea (102,103). In most of the studies exogenous arachidonate was added to stimulate hydroperoxide synthesis and cooxygenation. Recently, though, Amstad and Cerutti reported that the levels of aflatoxin B_1-DNA adducts formed in C3H 10T½ fibroblasts were decreased by treatment of the cells with indomethacin or eicosatetraynoic acid, inhibitors of arachidonate oxygenation (104). They concluded that a significant fraction of total aflatoxin epoxidation by 10T½ cells occurs as a result of arachidonate-dependent cooxygenation. This implies that cooxygenation takes place in cells and that it is triggered by release of arachidonate from endogenous stores.

Is it possible to quantitate the relative contribution of hydroperoxide-dependent and mixed-function oxidase-dependent oxidation of chemical carcinogens *in vitro*, in cells and organs, and *in vivo*? Adding arachidonic acid or NADPH to support oxidation *in vitro* gives a good estimate of oxidative potential but its relation to cellular oxidation is not straightforward. Likewise, "specific" inhibitors can be helpful in *in vitro* experiments but their use can be compromised in cellular, organismal, or *in vivo* experiments by overlapping specificities or altered potencies. For example, many compounds that inhibit lipoxygenase activity at low concentration in microsomal or cytoplasmic fractions are ineffective when they are employed in cellular experiments. The reason for the differential effect is unclear but the implication for the use of such compounds as *in vivo* inhibitors is obvious.

A potentially powerful probe for sorting out the contribution of hydroperoxide-dependent and mixed-function oxidase-dependent polycyclic hydrocarbon oxidation is stereochemistry. Figure 11 summarizes the stereochemical differences in epoxidation of (±)-BP-7,8-diol by hydroperoxide-dependent and mixed-function oxidase-dependent pathways. The (-)-enantiomer of BP-7,8-diol is converted primarily to the (+)-anti-diolepoxide by both pathways whereas the (+)-enantiomer of BP-7,8-diol is converted primarily to the (-)-anti-diolepoxide by hydroperoxide-dependent oxidation and to the (+)-syn-diolepoxide by mixed-function oxidases. The stereochemical course of oxidation by cytochrome P-450 isoenzymes was first elucidated for the 3-methylcholanthrene-inducible form but we have detected the same stereochemical profile using rat liver microsomes from control, phenobarbital-, or 3-methylcholanthrene-induced animals (60,85,86). The only differences between the microsomal preparations is the rate of oxidation.

The findings summarized in Figure 11 provide a practical diagnostic tool for distinguishing the two routes of oxidation. Reactions can be performed with cellular or subcellular preparations and (±)- or (+)-BP-7,8-diol; the tetraol hydrolysis products of the diolepoxides are separated by HPLC and quantitated (105). When the substrate is (±)-BP-7,8-diol an anti/syn ratio in excess of 2.5 is seen for peroxide-dependent oxidation and an anti/syn ratio of 1 for mixed-function oxidase-dependent oxidation. When the substrate is (+)-BP-7,8-diol the anti/syn ratio for the mixed-function oxidase-dependent reaction decreases to ~0.3 (105). The tenfold difference in the anti/syn ratio between peroxide- and cytochrome P-450-dependent epoxidation makes (+)-BP-7,8-diol an extremely sensitive indicator of the pathway of oxidation. We have exploited it to demonstrate that lipid peroxidation in rat liver microsomes causes epoxidation (97). By using (+)-BP-7,8-diol, we have been able to distinguish epoxidation caused by NADPH-dependent lipid peroxidation in 3-methylcholanthrene-induced rat liver microsomes (T. A. Dix, unpublished results). These microsomes contain an extremely active cytochrome P-450 toward BP-7,8-diol but it is possible to differentiate the contribution of lipid peroxidation to epoxidation by determining the yield of tetraols from the (-)-anti-diol-epoxide. Although it has been suspected for some time that lipid peroxidation could cause xenobiotic oxidation in the presence of an active cytochrome P-450, our studies of BP-7,8-diol oxidation provided the first clearcut demonstration of it. Stereochemistry has also been employed to detect arachidonic acid-dependent BP-7,8-diol epoxidation in cultured hamster trachea (103). These examples illustrate the power of such stereochemical probes.

To what extent does cooxygenation occur *in vivo* and is it important in chemical carcinogenesis? This is a very difficult question to answer at the present time. Recent results demonstrate that aromatic amines and diamines can be cooxidized *in vivo* (64). In the case of β-napthylamine it is estimated that 30% of the adducts that form to DNA in the dog bladder, a target organ for napthylamine carcinogenesis, arise as a result of

arachidonate-dependent cooxidation (64). This conclusion is
based on the detection of unique peroxidase adducts to DNA that
are structurally distinct from mixed-function oxidase-generated
adducts. In contrast, pretreatment of A/HeJ mice with aspirin or
indomethacin does not lower the level of DNA adducts formed from
BP in lung nor does it reduce the incidence of lung neoplasms
induced by BP (105). Control experiments indicate that aspirin
treatment abolishes PGH synthase activity *in vivo*. This suggests
that PGH synthase-dependent cooxidation does not play a role in
lung tumorigenesis by benzo[a]pyrene in the adenoma model. This
may be related to the high levels of the endogenous antioxidant,
vitamin E, in rodent lung (106). However, administration of
aspirin to guinea pigs does not lower the levels of protein or
DNA adducts formed from BP in several different tissues, so the
levels of vitamin E may not be a determinant of BP cooxidation
(107).

The tissue distribution of PGH synthase suggests that it
does not play a major role in systemic drug metabolism because
most of the tissues where it is present in high concentration do
not receive a significant proportion of cardiac output. However,
several tissues with high PGH synthase activity, e.g., kidney and
uterus, are target organs for carcinogens that require metabolic
activation. In order to detect arachidonate-dependent metabolic
activation in these tissues, it will be necessary to develop
unique and specific probes. If systemic metabolism of a given
compound proceeds with a unique pattern of stereochemistry (e.g.,
BP-7,8-diol) or produces unique DNA adducts (β-napthylamine) then
it should be possible to quantitate the extent to which unique
stereoisomers or DNA adducts are formed. Hopefully, by using
such diagnostic probes it will be possible to provide quantitative answers to the questions about the extent to which cooxidation occurs *in vivo*.

The emphasis of this chapter has been on the role cooxidation plays in carcinogenesis. However, because metabolism and
activation also play a key role in toxicity and teratogenicity,
this pathway of xenobiotic metabolism could be important in these

pathological responses as well (108,109). Furthermore, free radicals have been implicated in tumor promotion. The enzymes of oxygenation of arachidonate provide the major sources of hydroperoxy fatty acids in virtually all cells and hydroperoxy fatty acids serve as latent sources of alkoxyl and peroxyl free radicals. Thus, this pathway for oxidant generation may be important in tumor promotion as well as initiation.

Acknowledgements

The research summarized in this chapter was supported by grants from the American Cancer Society (BC244) and National Institutes of Health (GM23642, CA22206 and CA32506). L.J.M. is a recipient of a Faculty Research Award from the American Cancer Society (FRA243). Celeste Kipke and Susan Lyman provided invaluable assistance in its composition.

REFERENCES

1. Samuelsson B, Goldyne M, Granstrom E, Hamberg M, Hammarstrom S, Malmsten C: Prostaglandins and thromboxanes. Ann. Rev. Biochem. (47): 997-1029, 1978.
2. Samuelsson B: Leukotrienes-mediators of immediate hypersensitivity. Reactions and inflammation.Science (220): 568-575, 1983.
3. Anderson W, Crutchley DJ, Chaudhari A, Wilson AGE, Eling TE: Studies on the covalent binding of an intermediate(s) in prostaglandin biosynthesis to tissue macromolecules. Biochem. Biophys. Acta (573): 40-50, 1979.
4. Basu A, Marnett L.J: Unequivocal demonstration that malondialdehyde is a mutagen. Carcinogenesis (4): 331-333, 1983.
5. Salomon RG, Miller DB, Zagorski MG, Revison BS, Lal K, Raychaudhuri SR, Avasthi K: Levuglandins: Isolation, characterization, and total synthesis of new secoprostanoid products from prostaglandin endoperoxides. Abst. Kyoto Conf. on Prostaglandins, Abstract #S4-17.
6. Hamberg M, Svensson J, Wakabayashi T, Samuelsson B: Isolation and structure of 2 prostaglandin endoperoxides that cause platelet-aggregation. Proc. Natl. Acad. Sci. USA (71): 345-349, 1974.
7. Nugteren DH, Hazelhof E: Isolation and properties of intermediates in prostaglandin biosynthesis. Biochim. Biophys. Acta (326): 448-461, 1973.
8. Hamberg M, Samuelsson B: Oxygenation of unsaturated fatty acids by the vesicular gland of sheep. J. Biol. Chem. (242): 5344-5354, 1967.

9. Diczfalusy U, Falardeau P, Hammarstrom S: Conversion of prostaglandin endoperoxides to C_{17}-hydroxy acids catalyzed by human platelet thromboxane synthase. FEBS Letts. (84): 271-274, 1977.

10. Watanabe K, Yamamoto S, Hayaishi O: Reaction of prostaglandin endoperoxides with prostaglandin I synthetase solubilized from rabbit aorta microsomes. Biochem. Biophys. Res. Commun. (87): 192-199, 1979.

11. Bernheim F, Bernheim MLC, Wilbur KM: The reaction between thiobarbituric acid and the oxidation products of certain lipids. J. Biol. Chem. (174): 257-264, 1948.

12. Shamberger RJ, Andreone TL, Willis CE: Antioxidants and cancer. IV. Initiating activity of malonaldehyde as a carcinogen. J. Natl. Cancer Inst. (53): 1771-1773, 1974.

13. Mukai FH, Goldstein BD: Mutagenicity of malondialdehyde, a decomposition product of peroxidized polyunsaturated fatty acids. Science (191): 868-869, 1976.

14. Marnett LJ, Tuttle MA: Comparison of the mutagenicities of malondialdehyde and the side products formed during its chemical synthesis. Cancer Res. (40): 276-282, 1980.

15. Fischer SM, Olge S, Marnett LJ, Nesnow S, Slaga TJ: The lack of initiating and/or promoting activity of sodium malondialdehyde on Sencar mouse skin. Cancer Letts. (19): 61-66, 1983.

16. Apaja M: Evaluation of toxicity and carcinogenicity of malonaldehyde. Acta Universitatis Ouluensis, Series D. (55): 1-61, 1980.

17. Marnett LJ, Buck J, Tuttle MA, Basu AK, Bull AW: Distribution and oxidation of malondialdehyde in mice. Prostaglandins (In press).

18. Basu AK, Marnett LJ: Molecular requirements for the mutagenicity of malondialdehyde and related acroleins. Cancer Res. (44): 2848-2854, 1984.

19. Malaveille C, Bartsch H, Grover PL, Sims P: Mutagenicity of non-K-region diols and diol-epoxides of benz(a)anthracene and benzo(a)pyrene in S. typhimurium TA100. Biochem. Biophys. Res. Commun. (66):693-700, 1975.

20. Wood AW, Wislocki PG, Chang RL, Levin W, Lu AYH, Yagi H, Hernandez O, Jerina DM, Conney AH: Mutagenicity and cytotoxicity of benzo(a)pyrene benzo-ring epoxides. Cancer Res. (36): 3358-3366, 1976.

21. Gardner HW: Lipid enzymes: Lipases, lipoxygenases, and hydroperoxidases. In: MG Simic and M. Karel (eds.) Autoxidation in Food and Biological Systems. Plenum, New York, pp. 447-504, 1980.

22. Benedetti A, Comporti M, Esterbauer H: Identification of 4-hydroxynonenal as a cytotoxic product originating from the peroxidation of liver microsomal lipids. Biochim. Biophys. Acta (620): 281-296, 1980.

23. Eder E, Heschler D, Neudecker T: Mutagenic properties of allylic and α,β-unsaturated compounds: consideration of alkylating mechanisms. Xenobiotica (12): 831-848, 1982.

24. Marnett LJ, Hurd H, Hollstein M, Levin DE, Esterbauer H, Ames BN: Naturally-occurring carbonyl compounds are mutagens in Salmonella tester strain TA104. Mutat. Res. (In press).

25. Miller JA, Miller EC: The initiation stage of chemical car-
    cinogenesis: An introductory overview. In: (TJ Powles, RS
    Bockman, KV Honn and P Ramwell) Prostaglandins and Cancer:
    First International Conference. Alan R. Liss, Inc., New
    York, pp. 81-96.
26. Conney AH: Induction of microsomal-enzymes by foreign
    chemicals and carcinogenesis by polycyclic aromatic-hydro-
    carbons. Cancer Res. (42): 4875-4917, 1982.
27. Levin W, Lu AYH, Ryan D, Wood AW, Kapitulnik J, West S,
    Huang M-T, Conney AH, Thakker DR, Holder G, Yagi H, Jerina
    DM: Properties of liver microsomal monooxygenase system and
    epoxide hydrase-factors influencing metabolism and muta-
    genicity of benzo[a]pyrene. In: HH Hiatt, JD Watson and JA
    Winsten (eds.) Origins of Human Cancer. Cold Spring Harbor,
    Cold Spring, pp. 659-682, 1977.
28. Rydstrom J, Montelius J, Bengtsson M: In: Extrahepatic
    Drug Metabolism and Chemical Carcinogenesis. Elsevier, New
    York, 1983.
29. Cavalieri EL, Rogan EG: One-electron and two-electron
    oxidation in aromatic hydrocarbon carcinogenesis. In: WA
    Pryor (ed.) Free Radicals in Biology. Academic Press, New
    York, Vol. 6, pp. 323-369, 1984.
30. Bartsch H, Hecker E: Metabolic activation of carcinogen
    N-hydroxy-N-2-acetylaminofluorene-3. Oxidation with horse-
    radish-peroxidase to yield 2-nitrosofluorene and N-acetoxy-
    N-2-acetylaminofluorene. Biochim. Biophys. Acta (237):
    567-578, 1971.
31. Floyd RA, Soong LM, Culver PL: Cancer Res. (36):
    1510-1519, 1976.
32. Metzler M, McLachlan JA: Peroxidase-mediated oxidation: A
    possible pathway for metabolic activation of diethylstil-
    besterol. Biochem. Biophys. Res. Commun. (85): 874-884,
    1978.
33. Badwey JA, Karnovsky ML: Active oxygen species and the
    functions of phagocytic leukocytes. Ann. Rev. Biochem.
    (49): 695-726, 1980.
34. Tolbert NE: Metabolic pathways in peroxisomes and glyoxy-
    somes. Ann. Rev. Biochem. (49): 695-726, 1980.
35. White RE, Coon MJ: Oxygen activation by cytochrome P-450.
    Ann. Rev. Biochem. (49): 315-356, 1980.
36. Hamberg M, Samuelsson B. Prostaglandin endoperoxides-novel
    transformations of arachidonic acid in human platelets.
    Proc. Natl. Acad. Sci. USA (71): 3400-3404, 1974.
37. Bryant RW, Simon TC, Bailey JM: Role of glutathione per-
    oxidase and hexone-monophosphate shunt in platelet lipoxy-
    genase pathway. J. Biol. Chem. (257): 14937-14943, 1982.
38. Ohki S, Ogino N, Yamamoto S, Hayaishi O: Prostaglandin hydro-
    eroxidase, an integral part of prostaglandin endoperoxide
    synthetase from bovine vesicular gland microsomes. J. Biol.
    Chem. (254): 839-846, 1979.
39. Marnett LJ: Hydroperoxide-dependent oxidations during prosta-
    glandin biosynthesis. In: WA Pryor (ed.) Free Radicals in
    Biology, Vol. 6. Academic Press, New York, pp. 63-94,
    1984.

40. Marnett LJ, Eling TE: Cooxidation during prostaglandin bio-
    synthesis: A pathway for the metabolic activation of xeno-
    biotics. In: E Hodgson, JR Bend and R.M. Philpot (eds.)
    Reviews in Biochemical Toxicology, Vol. 5. Elsevier/North
    Holland, New York, pp. 135-172, 1983.
41. Marnett LJ: Polycyclic hydrocarbon oxidation during prosta-
    glandin biosynthesis. Life Sci. (29): 531-546, 1981.
42. Marnett LJ, Reed GA, Johnson JT: Prostaglandin synthase
    dependent benzo(a)pyrene oxidation: Products of the oxida-
    tion and inhibition of their formation by antioxidants.
    Biochem. Biophys. Res. Commun. (79): 569-576, 1977.
43. Lorentzen RJ, Caspary WJ, Lesko SA, Ts'o POP: Autoxidation
    of 6-hydroxybenzo[a]pyrene and 6-oxobenzo[a]pyrene radical,
    reactive metabolites of benzo[a]pyrene. Biochemistry (14):
    3970-3977, 1975.
44. Lesko S, Caspary W, Lorentzen R, Ts'o POP: Enzyme formation
    of 6-oxobenzo[a]pyrene radical in rat liver homogenates from
    carcinogenic benzo[a]pyrene. Biochemistry (14): 3978-3984,
    1975.
45. Marnett LJ, Reed GA: Peroxidatic oxidation of benzo[a]-
    pyrene and prostaglandin biosynthesis. Biochemistry (18):
    2923-2929, 1979.
46. Bickers DR, Mukhtar H, Dutta-Choudhury T, Marcelo CL,
    Voorhees JJ: Aryl hydrocarbon hydroxylase, epoxide
    hydrolase, and benzo[a]pyrene metabolism in human
    epidermis: Comparative studies in normal subjects and
    patients with psoriasis. J. Invest. Dermatol. (83):
    51-56, 1984.
47. Sivarajah K, Anderson MW, Eling T: Metabolism of benzo[a]-
    pyrene to reactive intermediate(s) via prostaglandin bio-
    synthesis. Life Sci. (23): 2571-2578, 1978.
48. Marnett LJ, Reed GA, Dennison DJ: Prostaglandin synthetase
    dependent activation of 7,8-dihydro-7,8-dihydroxy-benzo[a]-
    pyrene to mutagenic derivatives. Biochem. Biophys. Res.
    Commun. (82): 210-216, 1978.
49. Wislocki PG, Wood AW, Chang RL, Levin W, Yagi H, Hernandez
    O, Dansette PM, Jerina DM, AH Conney: Mutagenicity and
    cytotoxicity of benzo(a)pyrene arene oxides, phenols,
    quinones, and dihydrodiols in bacterial and mammalian cells.
    Cancer Res. (36): 3350-3357, 1976.
50. Fahl WE, Scarpelli D, Gill K: Association of specific
    chromosome abnormalities with type of acute leukemia and
    with patient age. Cancer Res. (41): 3400-3406, 1981.
51. Brookes P, Osborne MR: Mutation in mammalian cells by stereo-
    isomers of anti-benzo[a]pyrene diolepoxide in relation to
    the extent and nature of the DNA reaction-products. Car-
    cinogenesis (3): 1223-1226, 1982.
52. Adriaenssens PI, White CM, Anderson MW: Dose-response
    relationship for the binding of benzo(a)pyrene metabolites
    to DNA and protein in lung, liver, and forestomach of con-
    trol and butylated hydroxyanisole-tretated mice. Cancer
    Res. (43): 3712-3719, 1983.
53. Ashurst SW, Cohen GM, Nesnow S, DiGiovanni J, Slaga TJ:
    Formation of benzo(a)pyrene/DNA adducts and their relation-
    ship to tumor initiatin in mouse epidermis. Cancer Res.
    (43): 1024-1029, 1983.

54. Marnett LJ, Johnson JT, Bienkowski MJ: Arachidonic acid dependent metabolism of 7,8-dihydroxy-7,8-dihydro-benzo[a]-pyrene by ram seminal vesicles. FEBS Letts. (106): 13-16, 1979.

55. Sivarajah K, Mukhtar H, Eling T: Arachidonic acid-dependent metabolism of (±) trans-7,8-dihydroxy-7,8-dihydro-benzo[a]-pyrene (B,P-7,8-diol) to 7,10/8,9 tetrols. FEBS Letts. (106): 17-20, 1979.

56. Marnett LJ, Bienkowski MJ: Hydroperoxide-dependent oxygena-tion of 7,8-dihydroxy-7,8-diydrobenzo[a]pyrene by ram seminal vesicle microsomes. Source of the oxygen. Biochem. Biophys. Res. Commun. (96): 639-647, 1980.

57. Marnett LJ, Panthananickal A, Reed GA: Metabolic activation of 7,8-dihydroxy-7,8-dihydrobenzo[a]pyrene during prosta-glandin biosynthesis. Drug. Metab. Rev. (13):235-247, 1982.

58. Panthananickal A, Marnett LJ: Arachidonic acid-dependent metabolism of 7,8-dihydroxy-7,8-dihydrobenzo[a]pyrene to polyguanylic acid-binding derivatives. Chem. Biol. Interact. (258): 4411-4418, 1983.

59. Reed GA, Marnett LJ: Metabolism and activation of 7,8-dihydro-benzo[a]pyrene during prostaglandin biosynthesis: Intermediacy of a bay-region epoxide. J. Biol. Chem. (257): 11368-11376, 1982.

60. Panthananickal A, Weller P, Marnett LJ: Stereoselectivity of the epoxidation of 7,8-dihydrobenzo[a]pyrene by prosta-glandin H synthase and cytochrome P-450 determined by the. identification of polyguanylic acid adducts. J. Biol. Chem. (258): 4411-4418, 1983.

61. Guthrie J, Robertson IGC, Zeiger E, Boyd JA, Eling TE: Selective activation of some dihydrodiols of several poly-cyclic aromatic hydrocarbons to mutagenic products by prostaglandin synthetase. Cancer Res. (42): 1620-1623, 1982.

62. Lasker J, Sivarajah K, Mason RP, Kalyanaraman B, Abou-Donia MB, Eling TE: A free radical mechanism of prostaglandin synthase-dependent aminopyrine demethylation. J. Biol. Chem. (256): 7764-7767, 1981.

63. Morton KC, King CM, Vaught JB, Wang, CY, Lee MS, Marnett LJ: Prostaglandin H synthase-mediated reaction of carcinogenic arylamines with t-RNA and polynucleotides. Biochem. Biophys. Res. Commun. (111): 96-103, 1983.

64. Yamazoe Y, Miller DW, Gupta RC, Zenser TV, Weis CC, Kadlubar FF: DNA adducts formed by prostaglandin H synthase-mediated activation of carcinogenic arylamines. Proc. Amer. Assn. Cancer Res. (25): 91, 1984.

65. Siedlik PH, Marnett LJ: Oxidizing radical generation by prostaglandin synthase. Meth. Enzymol. (105): 412-416, 1984.

66. Marnett LJ, Bienkowski MJ, Pagels WR, Reed GA: Mechanism of xenobiotic cooxygenation coupled to prostaglandin H2 bio-synthesis. In: B Samuelsson, PW Ramwell and R Paoletti (eds.) Advances in Prostaglandins and Thromboxane Research, Vol. 6. Raven Press, New York, pp. 149-151, 1980.

67. Pryor WA: Free radicals in autoxidation and in aging. Part I. Kinetics of the autoxidatin of linoleic acid in SDS micelles; Calculations of radical concentrations, kinetic chain lengths, and the effects of Vitamin E. In: D Armstrong (ed.) Free Radicals in Biology and Aging. Raven Press, New York. (In press).

68. Reed GA, Brooks EA, Eling TE: Phenylbutazone-dependent epoxidatin of 7,8-dihdyroxy-7,8-dihydrobenzo(a)pyrene. J. Biol. Chem. (259): 5591-5595, 1984.

69. Samokyszyn VM, Sloane BF, Honn KV, Marnett LJ: Cooxidation of 13-cis retinoic acid by prostaglandin H synthase. Biochem. Biophys. Res. Commun. (124): 430-436, 1984.

70. Verma AK, Rice HM, Shapas BD, Boutwell RK: Inhibition of 12-0-tetradecanoylphorbol-13-acetate-induced ornithine decarboxylase activity in mouse epidermis by vitamin A analogs (retinoids). Cancer Res. (38): 793-801, 1979.

71. Hennings H, Wenk ML, Donahoe R: Retinoic acid promotion of papilloma formation in mouse skin. Cancer Letts. (16): 1-5, 1982.

72. Marnett LJ, Wlodawer P, Samuelsson B: Cooxidation of organic substrates by prostaglandin synthetase of sheep vesicular gland. J. Biol. Chem. (250): 8510-8517, 1975.

73. Sivarajah K, Lasker JM, Eling TE: Prostaglandin synthetase-dependent cooxidation of ($\pm$)-benzo[a]pyrene-7,8-dihydrodiol by human lung and other mammalian tissues. Cancer Res. (41): 1843-1847, 1982.

74. Zenser TV, Mattamal MB, Davis BB: Differential distribution of the mixed-function oxidase activities in rabbit kidney. J. Pharmacol. Exp. Ther. (207): 719-725, 1978.

75. Rapp NS, Zenser TV, Brown WW, Davis BB: Metabolism of benzidine by a prostaglandin-mediated process in inner renal medullary slices. J. Pharmacol. Exp. Ther. (215): 401-407, 1980.

76. Van der Ouderaa FJ, Buytenhek M, Nugteren DH, Van Dorp DA: Purification and characterization of prostaglandin endoperoxide synthetase from sheep vesicular glands. Biochim. Biophys. Acta (487): 315-331, 1977.

77. Van der Ouderaa FJ, Buytenhek M, Slikkerveer FJ, Van Dorp DA: On the haemoprotein character of prostaglandin endoperoxide synthetase. Biochim. Biophys. Acta (572): 29-42, 1979.

78. Pagels WR, Sachs RJ, Marnett LJ, Dewitt DL, Day JS, Smith WL: Immunochemical evidence for the involvement of prostaglandin H synthase in hydroperoxide-dependent oxidations by ram seminal vesicle microsomes. J. Biol. Chem. (258): 6517-6523, 1983.

79. Koshihara Y, Senshu T, Kawamwa M, Murota S: Sodium n-butyrate induces prostaglandin synthetase in mastocytoma P815 cells. Biochim. Biophys. Acta (617): 1253-1258, 1980.

80. Murota S, Morita I: Prostaglandin-synthesizing system in rat liver: Changes with aging and various stimuli. In: B. Samuelsson, P.W. Ramwell and R. Paoletti (eds.) Advances in Prostaglandin and Thrmboxane Research, Vol. 8. Raven Press, New York, pp. 1495-1506, 1980.

81. Sivarajah K, Jones KG, Fouts JR, Devereux T, Shirley JE,
     Eling TE: Prostaglandin synthetase and cytochrome P450-
     dependent metabolism of (±) benzo[a]pyrene 7,8-dihydrodiol
     by enriched population of rat Clara cells and alveolar type
     II cells. Cancer Res. (43): 2632-2636, 1983.
82. Sivarajah K, Lasker JM, Eling TE, Abou-Donia MB: Metabolism
     of N-alkyl compounds during the biosynthesis of
     prostaglandins. Mol. Pharmacol. 921): 133-141, 1982.
83. Smith PBW, Caldwell J, Smith RL, Horner MW, Houghton E, Mass
     MS: The disposition of phenylbutazone in the horse.
     Abstracts Ninth European Workshop on Drug Metabolism.
     Abstract #P28, 1984.
84. Hrycay EG, O'Brien PJ: Cytochrome P-450 as a microsomal
     peroxidase in steroid hydroperoxide reduction. Arch.
     Biochem. Biophys. (153): 480-494, 1972.
85. McCarthy M-B, White RE: Functional differences between per-
     oxidase compound I and the cytochrome P-450 reactive oxygen
     intermediate. J. Biol. Chem. (258): 9153-9158, 1983.
86. Thakker DR, Yagi H, Akagi H, Koreeda M, Lu AYH, Levin W,
     Wood AW, Conney AH, Jerina DM: Metabolism of benzo(a)-
     pyrene. VI. Stereoselective metabolism of benzo(a)pyrene
     and benzo(a)pyrene-7,8-dihydrodiol to diol epoxides. Chem.
     Biol. Interact. (16): 281-300, 1977.
87. Deutsch J, Leutz JC, Yang SK, Gelboin HV, Chang YL, Vatsis
     KP, Coon MJ: Regio-and stereoselectivity of various forms
     of cytochrome P450 in the metabolism of benzo(a)pyrene and
     (-) trans-7,8-dihydroxy-7,8 dihydrobenzo(a)pyrene as shown
     by product formation and binding to DNA. Proc. Natl. Acad.
     Sci. USA. (75): 3123-3127, 1978.
88. Dunford HB, Stillman JS: On the function and mechanism of
     action of peroxidases. Coord. Chem. Rev. (19): 187-251,
     1976.
89. Weller P, Markey CM, Marnett LJ: Enzymatic reduction of
     5-phenyl-4-pentenyl hydroperoxide: Detection of peroxidases
     and identification of peroxidase reducting substrates. Arch.
     Biochem. Biophys. (Submitted).
90. Dix TA, Marnett LJ: Free radical epoxidation of 7,8-
     dihydroxy-7,8-dihydrobenzo[a]pyrene by hematin and poly-
     unsaturated fatty acid hydroperoxides. J. Amer. Chem. Soc.
     (103): 6744-6746, 1981.
91. Dix TA, Fontana R, Panthani A, Marnett: J. Biol. Chem. (In
     press).
92. Morgenstern R, DePierre JW, Lind C, Guthenberg C, Mannervik
     B, Ernster L: Benzo(alpha) pyrene quinones can be generated
     by lipid peroxidation and are conjugated with glutathione by
     glutathione-S-transferase-B from rat liver. Biochem.
     Biophys. Res. Commun. (99): 682-690, 1981.
93. Gower JD, Wills ED: The generation of oxidation products of
     benzo[a]pyrene by lipid peroxidation: A study using
     γ-irradiation. Carcinogenesis (5):1183-1189, 1984.
94. Mayo FR: Free-radical autoxidations of hydrocarbons. Acc.
     Chem. Res. (1): 193-201, 1968.
95. Battista JR, Marnett LJ: Prostaglandin H synthase-dependent
     epoxidation of aflatoxin B_1. Carcinogenesis (Submitted).
96. Porter NA: Chemistry of lipid peroxidation. Meth. Enzymol.
     (105): 273-282, 1984.

97. Dix TA, Marnett LJ: Metabolism of polycyclic aromatic hydro-
     carbon derivatives to ultimate carcinogens during lipid per-
     oxidation. Science (221): 77-79, 1983.
98. Lands WEM, Samuelsson B: Phospholipid precursors of prosta-
     glandins. Biochim. Biophys. Acta (164): 426-429, 1968.
99. Vonkeman H, Van Dorp DA: The action of prostaglandin
     synthetase on 2-arachidonyl-lecithin. Biochim. Biophys.
     Acta (164): 430-432, 1968.
100. Galli C, Galli G, Porcellati G: In: Advances in Prosta-
     glandin and Thromboxane Research, Vol. 3. Raven Press, New
     York, 1978.
101. Fletcher R: Lipids of human myocardium. Lipids. (7):
     728-732, 1972.
102. Wong PK, Hampton MJ, Floyd RA: Evidence for lipoxygenase-
     peroxidase activation of N-hydroxy-2-acetylamino-fluorene by
     rat mammary gland parenchymal cells. In: T.J. Powles, R.S.
     Bockman, K.V. Honn and P.W. Ramwell (eds.). Alan R. Liss,
     New York, pp. 167-179, 1982.
103. Reed GA, Grafstrom RC, Krauss RS, Autrup H, Eling TE:
     Prostaglandin-H synthase-dependent cooxygenation of $(\pm)$-7,8-
     hydroxy-7,8-dihydrobenzo[a]pyrene in hamster trachea and
     human bronchus explants. Carcinogensis (5): 955-960,
     1984.
104. Amstad P, Cerutti P: DNA-binding of aflatoxin B-1 by cooxy-
     genation in mouse embryo fibroblasts C3H-10T½ cells. Biochem.
     Biophys. Res. Commun. (112): 1034-1040, 1983.
105. Adriaenssens PI, Sivarajah K, Boorman GA, Eling TE, Anderson
     MW: Effect of aspirin and indomethacin on the formation of
     benzo(a)pyrene-induced pulmonary adenomas and DNA adducts in
     A/HEJ mice. Cancer Res. (43):4762-4767, 1983.
106. Kornbrust DJ, Mavis RD: Relative susceptibility of micro-
     somes from lung, heart, liver, kidney, brain and testes to
     lipid peroxidation-correlation with vitamin E contents.
     Lipids (15): 315-322, 1980.
107. Garattini E, Coccia P, Romano M, Jiritano L, Noseda A,
     Salmona M: Prostaglandin endoperoxide synthetase and the
     activation of benzo(a)pyrene to reactive metabolites in
     vivo in Guinea pigs. Cancer Res. (44): 5150-5155,
     1984.
108. Witz G, Goldstein BD, Amoruso M, Stone DS, Troll W: Retinoid
     inhibition of superoxide anion radical production by polymor-
     phouclear leukocytes stimulted by tumor promoters. Biochem.
     Biophys. Res. Commun. (97): 883-888, 1980.
109. Slaga TJ, Klein-Szanto AJP, Triplett LL, Yotti LP, Trosko
     JE: Tumor-promoting activity of benzoyl peroxide. A widely
     used free radical generating compound. Science (213):
     1023-1025, 1981.

3

ARACHIDONIC ACID-DEPENDENT METABOLISM OF CHEMICAL CARCINOGENS
AND TOXICANTS
THOMAS E. ELING AND ROBERT S. KRAUSS

1. INTRODUCTION

Exposure to environmental chemicals is generally recognized as an important cause of human cancer (1).
Investigations on the mechanisms responsible for the initiation of carcinogenesis by chemicals indicate that most chemicals must be metabolized to exert their carcinogenic
effects. The central hypothesis underlying current thinking on
the induction of cancer, as originally proposed by the
Millers, is that chemical carcinogens are converted to
electrophilic metabolites which covalently link to nucleophilic cellular macromolecules (2). DNA is considered the critical nucleophilic target associated with the induction of
neoplasia (2). Studies on the metabolic activation of chemical carcinogens have focused primarily on two major classes of
chemicals, the polycyclic aromatic hydrocarbons (PAH) and the
aromatic amines. Prototypes of these two classes are
benzo(a)pyrene, and 2-acetylaminofluorene or 2-naphthylamine,
respectively.

Metabolic activation of benzo(a)pyrene (BP) has been
extensively investigated. Several excellent reviews have
recently been published (3-5), and thus this subject will be
only briefly described here. Activation of benzo(a)pyrene
(BP) requires, first, epoxidation across the 7 and 8 positions
to form (+)-BP-7,8 oxide. The resulting epoxide is hydrolyzed
by epoxide hydrase to (-)-trans-BP-7,8-dihydrodiol, which in
turn is oxidized to two diolepoxides; 7β, 8α-dihydroxy-9α,
10α-epoxy-7,8,9,10-tetrahydrobenzo(a)pyrene (anti-diolepoxide)
and 7β, 8α-dihydroxy-9β, 10β-epoxy-7,8,9,10-
tetrahydrobenzo(a)pyrene (syn-diolepoxide) (Figure 1).

L.J. Marnett (ed.), ARACHIDONIC ACID METABOLISM AND TUMOR INITIATION.
Copyright © 1985. Martinus Nijhoff Publishing, Boston. All rights reserved.

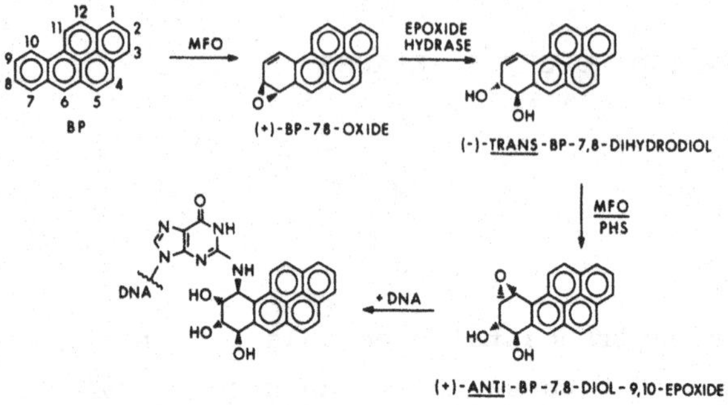

Figure 1.   Metabolic activation of benzo(a)pyrene.

Thus, BP-7,8-dihydrodiol is considered the proximate carcinogenic metabolite, while the diolepoxides are considered the
ultimate carcinogens.   The anti-diolepoxide is the major
isomer formed in vivo, but the ratio of anti to syn depends on
the tissue and species examined (3).   The anti isomer appears
to be more carcinogenic than the syn isomer (3).
BP-diolepoxides react with DNA, forming a linkage between C-10
on the polycyclic hydrocarbon and the exocyclic amino group of
guanine (6).   DNA adducts isolated from tissues or cells
treated with BP are predominantly those derived from the anti-
diolepoxide (4), with minor amounts of adducts derived from
the syn-diolepoxide detected.   The nature of the BP-DNA
interaction that causes a mutagenic or carcinogenic event is a
matter of considerable speculation.

The mechanism responsible for activation of carcinogenic
aromatic amines has also been intensively studied.   The importance of electrophilic metabolites in the initiation of carcinogenesis was derived initially from metabolic studies on
aromatic amines (2).   2-Acetylaminofluorene, 2-naphthylamine
and benzidine are members of this class of chemical carcinogens.   The currently accepted theory for the activation of
these liver and urinary bladder carcinogens involves formation
of an hydroxylamine metabolite (2,7).   Further metabolism to
"activated esters" is often required (2,7) (Figure 2).   The

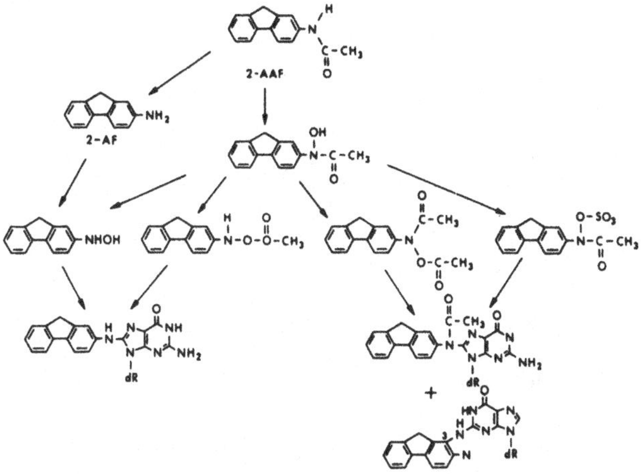

Figure 2.   Metabolic activation of 2-acetylaminofluorene.

hydroxylamine derivative is the proposed proximate carcinogen,
while a nitrenium ion formed either directly from the hydroxylamine or via the "activated ester" is considered the ultimate
carcinogen (2,7).   The hydroxylamine is considered to be an
obligatory intermediate in this activation scheme.   The major
DNA adducts isolated from the livers of animals treated with
these carcinogens are formed by linkage of the aromatic amine
nitrogen to C-8 of guanine (2,7,8).   Minor adducts are formed
with other purine bases, and in some cases, adducts formed
from the N-acetyl derivatives of the amines are also observed
(2,7,8).   The DNA adducts formed by aromatic amines in the
urinary bladder are not well characterized.

    The formation of electrophilic metabolites also appears to
be involved in the development of toxicities other than carcinogenesis.   For example, the widely used analgesic drug acetaminophen causes hepatic toxicity in high doses (9), and can
induce renal damage, even in moderate doses (10).   The
accepted mechanism for acetaminophen-induced toxicity is that
the drug is converted to an electrophilic metabolite(s) which

covalently binds to cellular proteins, producing necrosis.
The exact chemical nature of the electrophilic metabolite of
acetaminophen is uncertain (Figure 3), but evidence suggests

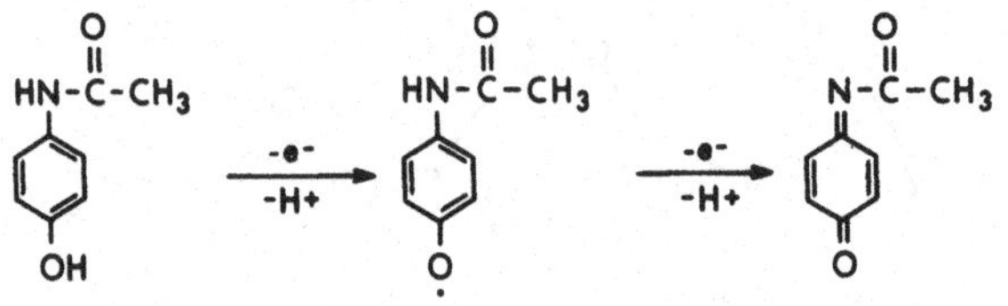

Figure 3.   Metabolic activation of acetaminophen.

that the phenoxy radical or N-acetyl-p-benzoquinonimine are
possible candidates (11,12).   Bromobenzene and 4-ipomeanol are
toxicants that also require metabolic activation (13,14).
These examples and others, indicate that ample evidence exists
for the involvement of electrophilic metabolites in the development of chemical toxicity.

An extensive literature testifies to the importance of
cytochrome P-450 monooxygenases in the metabolic activation of
chemicals to reactive electrophiles (4,15).   There is much
evidence, however, to suggest that the peroxidase activity of
prostaglandin H synthase can also metabolically activate
potentially toxic chemicals.   Considering the dissimilarity
between cytochrome P-450 and prostaglandin H synthase (PHS)
with respect to physical properties (15,16), response to inhibitors (17,18) and tissue distribution (15,19), it is possible
that PHS serves as an alternate enzyme for xenobiotic metabolism, particularly in tissues with low mixed-function oxidase
activity.   Consequently, our laboratory has investigated the
cooxidation of drugs and chemical carcinogens during
prostaglandin biosynthesis.

2. ARACHIDONIC ACID METABOLISM: A SOURCE OF PEROXIDES

Arachidonic acid is metabolized to hydroperoxy compounds by either prostaglandin H synthase (PHS) or specific lipoxygenases. PHS catalyzes two distinctly different enzymatic reactions (20,21). The cyclooxygenase activity of PHS converts arachidonic acid to the hydroperoxy endoperoxide PGG_2; PHS peroxidase reduces PGG_2 to the hydroxy endoperoxide, PGH_2. PGH_2 is further metabolized by additional enzymes to the classical prostaglandins (PGD_2, PGE_2, $PGF_{2\alpha}$), thromboxane A_2 and/or prostacyclin, depending on the tissue in which it is generated (Figure 4). Lipoxygenases metabolize arachidonic acid to 5-, 8-, 11- or 15-hydroperoxy fatty acids (HPETEs) that are reduced to the corresponding hydroxy fatty acids (22). The peroxidase responsible for the reduction of HPETEs may be PHS or glutathione peroxidase (23,24). Lipoxygenase specificity varies with the tissue and species. Lipoxygenase activity can often exceed the prostaglandin H synthase activity in some tissues. 5-HPETE can also be further metabolized to leukotrienes (22), a pathway that predominates in inflammatory cells (Figure 4). Nonsteroidal anti-inflammatory

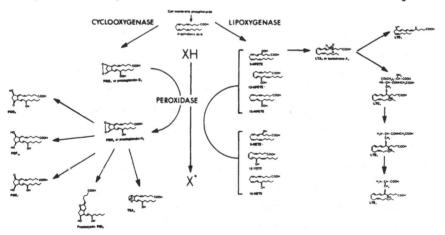

Figure 4. Metabolism of arachidonic acid to prostaglandins, leukotrienes and hydroperoxy fatty acids.

agents, for example aspirin or indomethacin, inhibit cyclooxygenase activity but not the peroxidase activity of PHS (25). No selective lipoxygenase inhibitor is available, but 5,8,11,14-eicosatetraynoïc acid and nordihydroguaiaretic acid

can inhibit both cyclooxygenase and lipoxygenase, however at
different concentrations (26).

The level of free arachidonic acid in most cells is low,
and thus PHS activity in cells is regulated in part by the
release of arachidonic acid from phospholipids by phospholipases. The release of arachidonic acid from phospholipids and
subsequent prostaglandin and hydroxy fatty acid formation
occurs in response to a wide variety of chemical and physical
stimuli. The peptide hormone, bradykinin, and the calcium
ionophore, A23187, stimulate the release of arachidonic acid
in endothelial cells (27), while thrombin and collagen stimulate phospholipase activity in platelets (28). The tumor promoter 12-0-tetradecanoylphorbol-13-acetate, causes arachidonic
acid release in a number of cells and tissues (29).
Mechanical stimulation, including perfusion or inhalation of
particulates, or altered respiration rate, increases arachidonic acid metabolism by the isolated perfused lung (30,31).
Cell division also alters arachidonic acid metabolism in cell
culture. Many growing cells produce significant amounts of
prostaglandins, while confluent monolayers of these cells produce only low amounts of arachidonic acid metabolites (32).

Anti-inflammatory steroids, such as hydrocortisone, dexa-
methasone, and betamethasone, inhibit phospholipase
A_2 activity and thus inhibit the release of arachidonic acid
(33). Thus the physiological or pathological state of the
tissue or the presence of xenobiotics can alter arachidonate
metabolism and hence alter cooxidation of toxicants and carcinogens by PHS.

Metabolism of chemicals is catalyzed by the peroxidase of
PHS. To catalyze the reduction of PGG_2 or HPETEs, PHS peroxidase requires reducing cofactors. These cofactors donate
single electrons to the peroxidase, and in turn are converted
to electron deficient metabolites which can be chemically
reactive. Ram seminal vesicle microsomes, a rich source of
PHS, in the presence of arachidonic acid will oxidize a wide
variety of compounds. Cooxidation of chemicals has also been

shown using other extrahepatic tissues possessing significantly less PHS activity than do ram seminal vesicles (34,53).

3. XENOBIOTICS METABOLIZED BY PHS: TYPES OF REACTIONS

A summary of the xenobiotics metabolized by PHS, and the
products of these oxidations, is presented in Table 1. This
illustrates the large number and diverse classes of chemicals
metabolized. Polycyclic aromatic hydrocarbons are poor
substrates for PHS and are not reducing cofactors for the
peroxidase. Metabolism of these carcinogens is dependent on a
hydroperoxy fatty acid and is not supported by H_2O_2 (93). In
contrast, aromatic amines and phenols are excellent substrates
for PHS and are reducing cofactors for PHS peroxidase.
Metabolism of these chemicals by PHS is supported by either
H_2O_2 or hydroperoxy fatty acids derived from arachidonic acid.
Aliphatic amines like benzphetamine are poor substrates. A
number of classical cytochrome P-450 substrates, e.g.,
7-ethoxycoumarin and 7-ethoxyresorufin, are very poor substa-
tes for PHS. However, like cytochrome P-450, the type of chemical reactions that are catalyzed by PHS is very diverse.
Dehydrogenations, N-demethylations, N-oxidation, and C-
oxidations occur and most likely proceed via a one electron
oxidation mechanism. Epoxidation of polycyclic aromatic
hydrocarbons like BP-7,8-dihydrodiol appears to be mediated by
peroxyl radicals. (See below).

3.1. N-Demethylation

Mono- and dimethyl aromatic amines are N-demethylated by
PHS (34).

$$AR-NH-CH_3 \xrightarrow[\text{ROOH}]{\text{PHS}} AR-NH_2 + HCHO$$

As seen in Table 1, a large number of N-methylated aromatic
amines have been studied. The reaction probably preceeds via
a free radical mechanism such as that described below for the
model compound aminopyrine, and appears to be different than
cytochrome P-450 mediated N-demethylation.

3.2. Sulfoxidation

Methyl phenyl sulfide and sulindac sulfide (35,36) are converted to the corresponding sulfoxides by PHS.

$$AR-S-CH_3 \longrightarrow AR-\overset{\overset{\displaystyle O}{\|}}{S}-CH_3$$

3.3. Dehydrogenation

Dehydrogenation is a characteristic reaction for peroxidases. Catechols (i.e., epinephrine) (37), and phenidone (38) both undergo dehydrogenation catalyzed by PHS. Likewise, the transplacental carcinogen diethylstilbestrol (DES) is oxidized to the corresponding semiquinone and subsequently to the quinone which rearranges to β-dienestrol (39) (Figure 5).

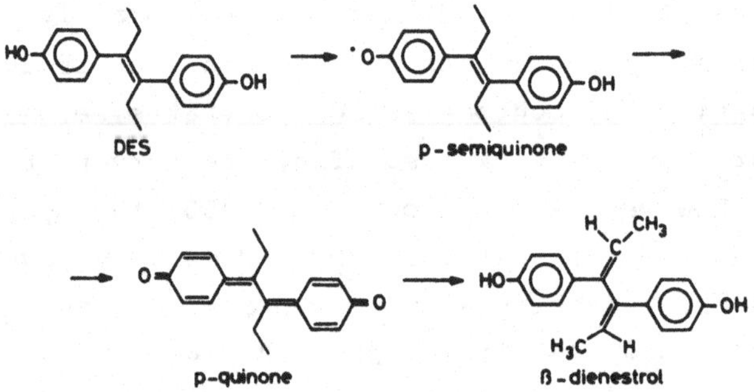

Figure 5. Metabolism of DES by PHS.

This pathway has been associated with metabolic activation of DES in vitro (40).

3.4. C-Oxidations

PHS can catalyze oxidations at carbon centers. For example, benzo(a)pyrene is metabolized by PHS to quinones (41,42). These quinones are presumed to arise by non-enzymatic decomposition of 6-hydroxy-BP (Figure 6).

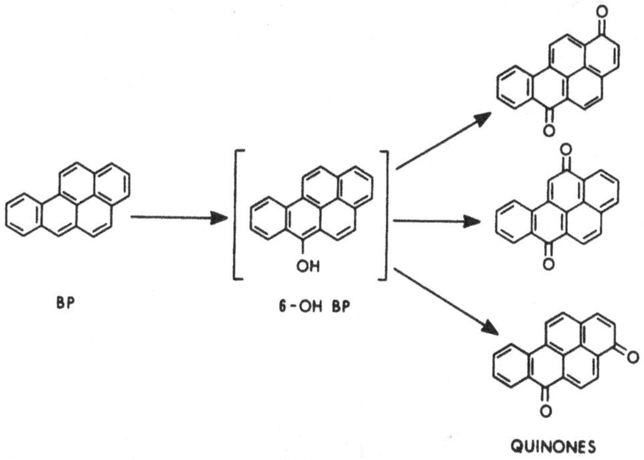

Figure 6.   Conversion of benzo(a)pyrene to benzo(a)pyrene
            quinones by PHS.

Phenylbutazone is also metabolized by PHS to a carbon-centered
free radical, which then traps oxygen, ultimately forming
4-hydroxyphenylbutazone (43).

3.5. N-Oxidation

Primary aromatic amines are excellent substrates for PHS
and undergo N-oxidation. Boyd and Eling have studied the
metabolism of 2-aminofluorene, a bladder carcinogen, by PHS
(44). Two of the isolable metabolites are 2,2'-azobisfluroene
and 2-nitrofluorene (Figure 7).

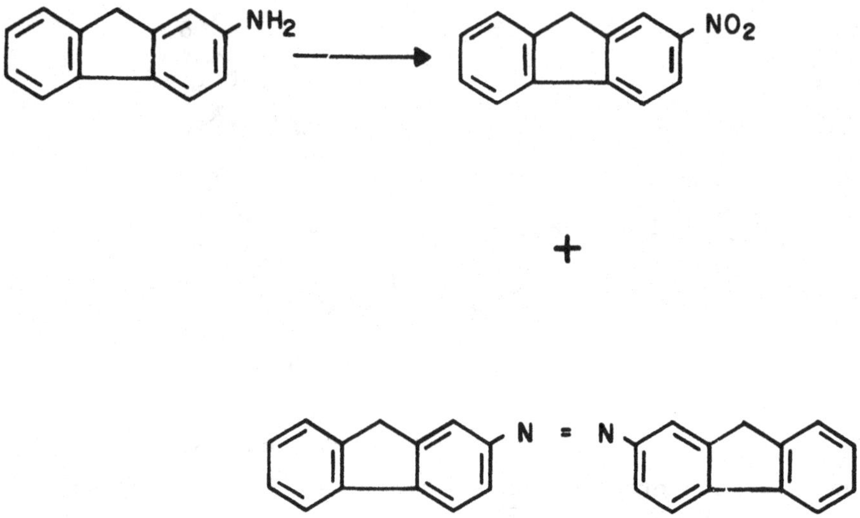

Figure 7. Metabolism of 2-aminofluorene by PHS.

4. FORMATION OF MUTAGENS BY PHS

Ample evidence exists for the formation of electrophilic metabolites from aromatic amines and polycyclic aromatic hydrocarbons by PHS. PHS used in place of the mixed-function oxidase in rat liver microsomes, oxidizes promutagens to mutagens as measured in strains of Salmonella typhimurium. As seen in Table 2, PHS selectively activates dihydrodiol metabolites of polycyclic aromatic hydrocarbons which contain an isolated double bond adjacent to the bay region to mutagens (45-47). The parent hydrocarbons and other diol metabolites are not activated. Thus, the proximate carcinogenic metabolites of polycyclic hydrocarbons are activated by PHS to ultimate carcinogenic forms.

Aromatic amines are also oxidized by PHS to derivatives mutagenic to S. typhimurium TA98 (48). 2-Aminofluorene, benzidine and 2-naphthylamine are activated to mutagens by PHS whereas 2-acetylaminofluorene and 1-naphthylamine are not (Table 2). 2-Nitrofluorene is one of the metabo-lites of PHS-dependent metabolism of 2-aminofluorene (44), thus it is possible that bacterial nitroreductase could reduce the nitro group to form 2-nitrosofluorene which is a mutagen in this strain. However, this does not appear to be the case since PHS-dependent mutagenicity of 2-aminofluorene is also observed with a bacterial strain deficient in nitroreductase (Boyd JA, Zeiger E, Eling TE, unpublished observations). These and other data suggest that a 2-aminofluorene free radical is the likely mutagenic metabolite. Rahimtula et al. recently reported that PHS metabolizes the aromatic amines 2-naphthylamine, p-phenetidine, and p-aminophenol to intermediates which produce DNA-strand breaks in cultured human fibroblasts (49). These results suggest that PHS is capable of activating a wide range of chemicals and may serve as an additional activating system to the mixed-function oxidases for studying or screening chemicals as potential mutagens.

5. MECHANISMS OF COOXIDATION BY PHS

The peroxidase activity of PHS catalyzes the oxidation of chemicals in a manner similar to the classical reactions catalyzed by peroxidases, that is reduction of a hydroperoxide occurs in the presence of an electron donor. PHS peroxidase may act mechanistically in a fashion very similar to horseradish peroxidase, although analogous enzyme intermediates have not yet been observed with PHS. Based on studies of oxidations catalyzed by horseradish peroxidase (HRP) the following model (Figure 8) is proposed. This model provides a useful tool for understanding PHS peroxidase-mediated oxidation of chemicals. The peroxide or hydroperoxide obtained from arachidonic acid oxidation, is reduced by the enzyme to the corresponding alcohol. During this process, the resting enzyme which contains a Fe^{+3} heme iron is oxidized to form an iron-oxo complex (analogous to HRP-compound I) in which the heme moiety is at a formal oxidation of +5 and the oxygen is derived from the hydroperoxide. "Compound I" can transfer the oxygen directly to an acceptor (A), regenerating the resting enzyme or be reduced by a single electron to an iron-oxo complex (analogous to HRP compound II) in which the heme iron is Fe^{+4}. An electron donor (DH) supplies the electron to the enzyme and thus becomes electron deficient. "Compound II" is then reduced to the resting state of the enzyme by the addition of a second electron supplied by another electron donor. Thus, the enzyme can oxidize a chemical by a one electron mechanism or by direct transfer of the hydroperoxide oxygen as noted in Figure 8. The chemistry of the electron deficient

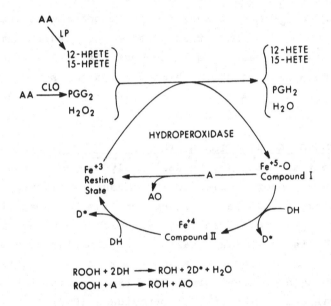

Figure 8. A hypothetical model of PHS catalyzed oxidations.

species will determine the product produced by a peroxidase-catalyzed reaction.

Sulindac sulfide is an example of a chemical oxidized by PHS peroxidase by direct oxygen transfer reactions. Egan and co-workers have shown that sulindac sulfide is oxidized to sulindac by PHS (35). The incorporated oxygen in sulindac is derived from the hydroperoxide oxygen as shown by O^{18} labeling studies (35). The stoichiometry between the reduction of 15-hydroperoxy-PGE_2 and oxidation of sulindac sulfide was approximately 1:1 (35). These data are in agreement with the oxygen transfer mechanism (Fig. 9).

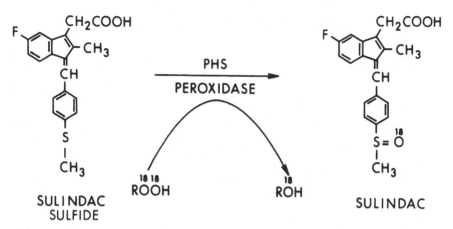

Figure 9. Oxidation of sulindac sulfide by PHS (35).

Most other peroxidase cosubstrates appear to be oxidized by the electron transfer reaction. We have studied the oxidative N-demethylation of aminopyrine by PHS peroxidase (50). The initial oxidation product is the aminopyrine free radical which can be easily measured due to its strong absorbance at 565 nm. Recently, we have examined the PHS peroxidase-catalyzed oxidation of aminopyrine and reduction of 15-HPETE (Eling TE, Sivarajah K, Mason RP, unpublished observations). A comparison of the 15-HPETE reduction to aminopyrene oxidized indicates a stoichiometry of 1 : 2 as shown in Table 3. These data suggest that amino-pyrene is oxidized by the electron transfer mechanism previously described in which 2 molecules of an electron donor are oxidized by one electron for every molecule of peroxide reduced.

Table 3

Comparison of 15-HPETE reduction to

aminopyrine oxidation by PHS.

nmoles of 15-HPETE Reduced	nmoles of AP Oxidized	$\dfrac{AP^{+\cdot}}{15\text{-HETE}}$
60	120	2.02
65	145	2.23
77	142	1.86

The reduction of 15-HPETE to 15-HETE was measured by HPLC and the aminopyrine free radical ($AP^{+\cdot}$) by measuring the absorbance at 565 nm.

It is clear that the peroxidase activity of PHS can catalyze several different kinds of hydroperoxide-dependent reactions including oxidation and oxygenation. Furthermore, in oxygenation reactions, the oxygen can be derived from molecular oxygen, hydroperoxide oxygen, or water (35,43,50,89,91).

The electron transfer mechanism described here applies to compounds which are reducing cofactors for PHS peroxidase. In contrast, several compounds which are oxygenated by PHS, including BP and BP-7,8-dihydrodiol, are not cofactors for the enzymatic reduction of hydroperoxides to alcohols, nor will simple hydroperoxides (i.e., hydrogen peroxide) support their oxygenation. This implies that these compounds are oxidized by unique mechanisms and that the oxidizing agent(s) is distinct from those functioning in peroxidase turnover.

Marnett and co-workers have extensively investigated the epoxidation of BP-7,8-dihydrodiol using a model system of hematin, detergent and unsaturated fatty acid hydroperoxides (51). The products and stereochemistry of the epoxidations by this model system are very similar to those observed for PHS. The epoxide oxygen was derived from molecular oxygen and the reaction was inhibited by antioxidants. A mechanism was proposed by Dix and Marnett which adequately explains the experimental observations seen in the model system (Figure 10). A

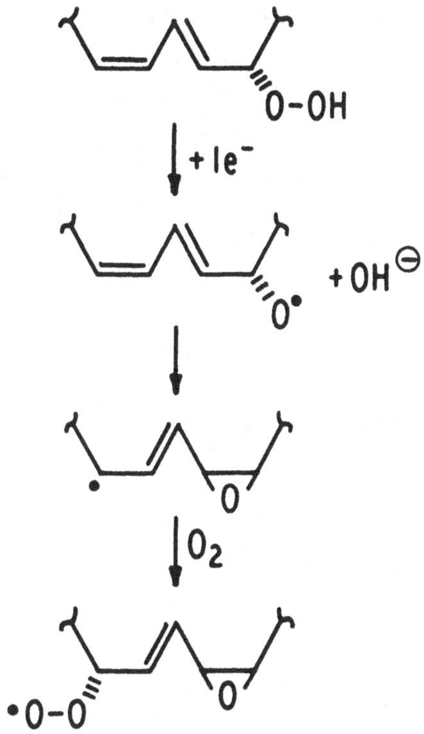

Figure 10. Proposed mechanism for the epoxidation of BP-7,8-
dihydrodiol by the hematin model system (51).

peroxyl radical formed from the unsaturated fatty acid is the
epoxidizing agent that converts BP-7,8-dihydrodiol to the
anti-diolepoxide. It is tempting to speculate that
PHS-dependent epoxidations proceed by a similar mechanism but
additional experimental evidence is needed to support this
conclusion.

We have recently shown that the addition of the anti-
inflammatory drug phenylbutazone greatly enhances the epoxidation of BP-7,8-dihydrodiol by either arachidonic acid or
H_2O_2 supplemented ram seminal vesicle microsomes (52). In the
absence of phenylbutazone, only arachidonic acid-fortified PHS
catalyzed the epoxidation of BP-7,8-dihydrodiol, as expected.

However, the addition of phenylbutazone to the PHS system produced a concentration dependent increase in epoxidation using either arachidonic acid or H_2O_2 as cofactor. Thus both the extent of epoxidation and the specificity for the peroxide was altered by the addition of phenylbutazone. The <u>anti</u>-diolepoxide was the major metabolite, and the same stereochemistry was observed as with PHS or the hematin model. Moreover, other data indicates that phenylbutazone was metabolized and appeared to participate in the epoxidation of the BP-7,8-dihydrodiol (52) (Figure 11). PHS peroxidase metabo-

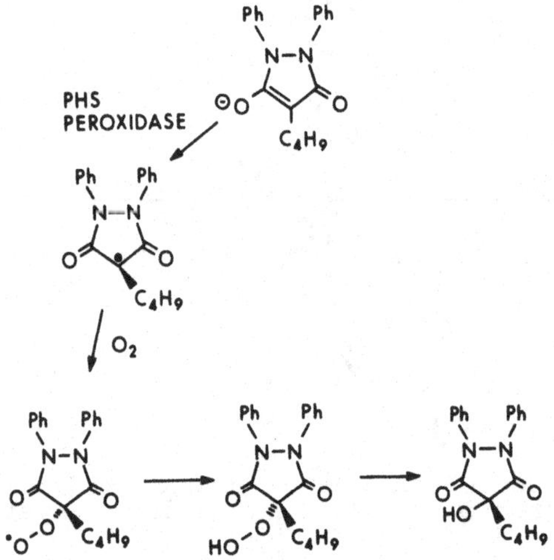

Figure 11. Metabolism of phenylbutazone by PHS (43).

lized phenylbutazone to 4-hydroxyphenylbutazone (43). Phenylbutazone donates an electron to the enzyme, forming a carbon centered radical at the 4 position. The radical traps molecular oxygen forming a peroxyl radical which is further converted to 4-hydroperoxyphenylbutazone. The hydroperoxy group is reduced to form 4-hydroxyphenylbutazone. Reed et al. (52) have proposed that in incubations containing phenylbutazone, the peroxyl radical of phenylbutazone is the epoxidizing agent that metabolizes BP-7,8-dihydrodiol to <u>anti</u>-diolepoxide.

This is analogous to a peroxyl radical formed by the hematin catalyzed decomposition of an unsaturated fatty acid hydroperoxide. In view of these findings an additional mechansim for the oxidation of xenobiotics by PHS is indicated, in which the oxidizing agent is derived from a primary peroxidase substrate. This mechanism should not be limited to phenylbutazone, but should occur with other peroxidase substrates which are metabolized to peroxyl radical intermediates as well. Although phenylbutazone is a reducing cofactor for PHS, oxidation of phenylbutazone by this mechanism serves to amplify the oxidative capacity of this enzyme rather than reduce it.

6. METABOLISM OF POLYCYCLIC AROMATIC HYDROCARBONS

PAH are a class of compounds that have been extensively studied as cooxidation substrates. Many PAH are carcinogenic (3-5). Oxidative metabolism of PAH has important biological consequences, since the metabolites are the actual carcinogens rather than the parent hydrocarbons. Benzo(a)pyrene (BP) is used as a model compound for the PAH.

BP was metabolized by ram seminal vesicle preparations following the addition of the PHS substrate arachidonic acid (41,42). The three stable products of BP cooxidation were the 1,6-, 3,6-, and 6,12-quinones (Figure 6), as determined by visible spectroscopy and cochromatography by TLC and HPLC with authentic standards (41,42). Metabolism of BP was dependent on the addition of arachidonic acid and was inhibited by indomethacin, establishing a role for PHS in the pathway. BP-cooxidation was supported by PGG_2 but not by PGH_2, indicating the peroxidative nature of the metabolism. Arachidonic acid, PGG_2, and the fatty acid hydroperoxide 15-HPETE were equivalent substrates for initiating BP cooxidation, but t-butyl and cumene hydroperoxides were not active (93). The cytochrome P-450 monooxygenases are not involved in BP cooxidation in ram seminal vesicle preparations, since cytochrome P-450 is not detectable spectrally in ram seminal vesicle microsomes nor was NADPH-dependent oxidation of BP (41,42).

Thus, this investigation of cooxidation during PG biosynthesis showed that the stable metabolites of BP cooxidation were the 1,6-, 3,6-, and 6,12-quinones, generally considered detoxication products. Considerable evidence has accumulated that the ultimate carcinogenic metabolites of BP are the bay region diolepoxides (3-5), thus subsequent studies have centered on cooxidation of the proximate carcinogen BP-7,8-dihydrodiol. (±)BP-7,8-dihydrodiol is epoxidized by PHS in the presence of arachidonic acid or hydroperoxyfatty acids to (±)-anti-diolepoxide (17,18). Little or no syn-diolepoxide is formed. The stereochemistry of epoxidation of BP-7,8-dihydrodiol by cooxidation is distinctly different from the stereochemistry of epoxidation of this compound by cytochrome P-450, in that both (+) and (-) BP-7,8-dihydrodiols are converted by PHS to anti-diolepoxides (95) while the cytochrome P-450 monooxygenases convert the (+) isomer to the syn-diolepoxide, and the (-) isomer to the antidiolepoxide (3-5) (Figure 12).

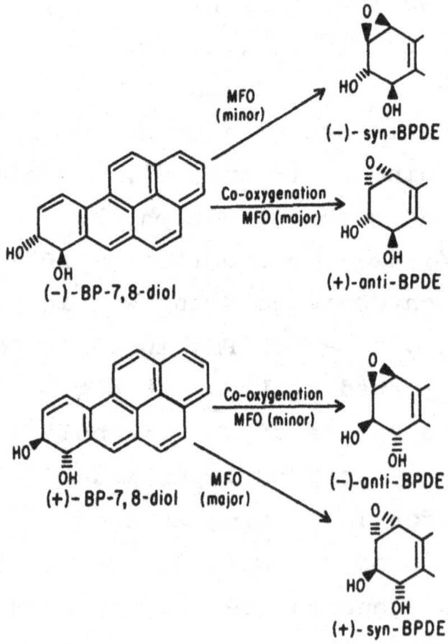

Figure 12. Stereochemical epoxidation of (+)-BP-7,8-dihydrodiol by PHS and mixed function oxidase.

Arachidonic acid-dependent BP-7,8-dihydrodiol cooxidation occurs in microsomal systems derived from a variety of tissues and species. Ram seminal vesicle, guinea pig lung, mouse skin, and rabbit kidney preparations produced the highest activity, while rat lung and intestine and human lung also cooxidized BP-7,8-dihydrodiol, though to a lesser extent (53). In all cases, anti-diolepoxide was the major product, and arachidonic acid-dependent epoxidation was inhibited by indomethacin. A comparison of arachidonic acid and NADPH-dependent metabolism of BP-7,8-dihydrodiol by guinea pig and human lung microsomes showed equivalent rates and total extents of BP-7,8-dihydrodiol metabolism regardless of whether NADPH or arachidonic acid was added (53).

Studies on the metabolism of PAH by PHS in in vitro systems has clearly shown that the proximate carcinogenic metabolites, the dihydrodiols, are converted to the diolepoxides. It is important to note that the reaction was dependent on hydroperoxide and PHS acted to generate its own source of peroxide (i.e. PGG_2). However, unsaturated fatty acid hydroperoxides derived from other sources may support this reaction as well. This point is important in designing experiments attempting to detect peroxidase-mediated metabolism in cell culture and in vivo, an area toward which future investigations must proceed.

Boyd et al. studied the metabolism of (±)-BP-7,8-dihydrodiol in the mouse embryo fibroblast cell line, C3H 10T½ clone 8 (54). Addition of arachidonic acid to confluent monolayers of cells stimulated metabolism of (±)-BP-7,8-dihydrodiol to the anti-diolepoxide by 2-3 fold; this stimulation was inhibited by the cyclooxygenase inhibitor indomethacin. Although basal metabolism of (±)-BP-7,8-dihydrodiol was unaffected by indomethacin, it has been previously shown that basal PG biosynthesis is low in confluent monolayers of these fibroblasts (32). The addition of arachidonic acid to cultures increased the frequency of transformation of these cells by (±)-BP-7,8-dihydrodiol and this

increase was inhibited by concomitant incubation with indomethacin (54). Similar studies with enriched populations of
rat non-ciliated bronchial (Clara) cells and alveolar type II
cells were performed with (±)-BP-7,8-dihydrodiol (55).
Utilizing exogenous addition of NADPH or arachidonic acid,
type II cells had greater capacity for PHS-dependent epoxidation of (±)-BP-7,8-dihydrodiol, whereas Clara cells had
greater capacity for monoxygenase-dependent epoxidation (Table
4).

Table 4

Comparison of PHS to cytochrome P-450 dependent
metabolism of (±)-benzo(a)pyrene
7,8-dihydrodiol by enriched rat lung cells.

Cell Preparation	PHS	P-450	Ratio P-450 PHS
Enriched Clara Cells	9.2 ± 1.1	21.6 ± 2.7	2.3
Enriched Type II Cells	8.2 + 2.0	5.1 ± 1.0	0.58

Values are mean ± S.D. N = 3   pmole of BP tetraols/min/10^6
cells.

Reed et al. have recently studied the role of PHS in the
metabolism of BP-7,8-dihydrodiol in explant cultures of
hamster and human tracheobronchial tissue (56). The addition
of arachidonic acid to hamster tracheas caused a significant
increase in the formation of anti-diolepoxide. This stimulation was inhibited to basal levels by indomethacin, supporting
the role of PHS in this response. Increased formation of
diolepoxide-DNA adducts was also observed in the presence of
arachidonic acid but these increases did not correlate well
with the increases in anti-diolepoxide, as measured by formation of BP tetraols. With human bronchial explants, the
addition of arachidonic acid also produced an increased formation of anti-diolepoxide. However, greater variability in
metabolism and responses to arachidonic acid addition was
observed. The further addition of indomethacin to human

bronchial explants did not inhibit the arachidonic acid-dependent stimulation of metabolism as expected. Therefore, metabolism of labeled arachidonic acid was examined in both hamster trachea and human bronchial explants. Hamster trachea metabolized arachidonic acid to exclusively PHS metabolites whereas human bronchial explants produced predominantly lipoxygenase products (56). Lipoxygenases are not inhibited by indomethacin, and these data suggest that hydroperoxides produced by lipoxygenase, rather than PGG_2 formed by PHS, supported the metabolism of BP-7,8-dihydrodiol in human tissue.

Inhibitors of arachidonic acid metabolism have been used in assessing the role of cooxidation of BP in vivo, as well as in cell cultures. Adriaenssens, et al. demonstrated that aspirin treatment had no effect on BP-induced pulmonary adenomas in mice, or on BP metabolite/DNA adduct formation (57). Recently, we have extended this study to examine the effect of aspirin on the types of tumor which developed (Eling TE, Sivarajah, K, unpublished observation). Tumors in mice produced by BP treatment appear to arise from either Clara or Type II cells. Since the ratio of PHS to P-450-dependent metabolism is different in these cells (Table 4), aspirin treatment could possibly shift the tumor from type II cells which are rich in PHS to Clara cells which are rich in P-450. As seen in Table 5 aspirin treatment did not alter either the number of tumors nor the type of tumor produced by BP. These studies were complicated by the fact that metabolism of BP by cooxidation is a detoxication pathway, but cooxidation of the proximate carcinogen BP-7,8-dihydrodiol is an activation pathway.

Table 5

Effect of Aspirin on BP-Induced Pulmonary Tumors

Treatment	Tumors/Mouse ($x \pm$ S.E.)	Clara Cell[c] Type II
BP[a] (3.0 mg)	20.9 ± 2.1 N = 18	$\frac{10}{215}$
BP + ASA[b] (3.0 mg)	20.9 ± 2.1 N = 21	$\frac{5}{232}$
BP (1.5 mg)	3.7 ± 0.7 N = 25	$\frac{2}{55}$
BP + ASA (1.5 mg)	4.2 ± 0.7 N = 24	$\frac{5}{63}$
ASA	0.3 ± 0.1 N = 20	---
Control	0.5 ± 0.2 N = 25	---

[a]BP oral dose dissolved in corn oil. In two doses (1.5 mg each) 2-3 times apart.

[b]Aspirin (ASA) mg/kg i.p. twice a day mice treated 1 day before BP dose, during BP dosing and 1 day after BP dose.

[c]As measured by histology sections. Not all tumors could be examined.

The use of inhibitors of arachidonic acid metabolism may be problematic however, when applied to such in vivo experiments, for three reasons. First, cyclooxygenase inhibitors only inhibit cooxidation reactions in which cyclooxygenase is the source of hydroperoxide. In tissues where lipoxygenases or lipid peroxidation are the major sources of peroxides, these inhibitors would have no effect on peroxidase-mediated metabolism, and their use could lead to erroneous conclusions. Second, in experiments utilizing tumorigenicity as an endpoint, the role of arachidonic acid metabolism in later stages of carcinogenesis must be taken into account. Prostaglandins stimulate or inhibit cell division in several cell culture systems (58-61). Therefore, the effects of inhibitors of arachidonic acid metabolism on fixation of carcinogen induced damage to DNA (via cell division) could complicate the interpretation of results. In addition, an important role for both cyclooxygenase and lipoxygenase products in tumor promotion has been implicated (62,63). Lastly, the inhibitor may alter the distribution and metabolism of the carcinogen. In conclusion, the results of in vivo experiments that attempt to assess arachidonic acid-dependent metabolism of xenobiotics through the use of inhibitors of arachidonic acid metabolism are very difficult to accurately interpret, whether such results are positive or negative. Thus, the use of inhibitors of arachidonic acid metabolism does not appear to be a fruitful approach to assess the role of cooxidation in vivo.

A more suitable approach may be through the use of stable biochemical markers or endpoints for assessing peroxidase-mediated metabolism. Such markers have been obtained from studies on the mechanisms involved in cooxidation, and the identification of metabolites of chemicals produced by peroxidases. Peroxidases can produced unique metabolites and these metabolites should serve as a useful index for the pathway.

Stereoselective oxidation by peroxidases may serve as a tool for measuring peroxidase-mediated metabolism of (±)-BP-7,8-dihydrodiol in the presence of other enzyme systems. In this case, (-)-BP-7,8-dihydrodiol is oxidized by PHS and monoxygenases to the anti-diolepoxide. However, (+)-BP-7,8-dihydrodiol is oxidized by PHS to the anti-diolepoxide, while monooxygenases metabolise this compound to the syn-diolepoxide. Thus, the ratio of anti- to syn-diolepoxides is a measurement of the ratio of peroxide-dependent to monooxygenase-dependent metabolism. Dix and Marnett have used this technique in determining the ratio of monoxygenase- to lipid hydroperoxide-dependent epoxidation of (±)-BP-7,8-dihydrodiol in rat liver microsomes, and to demonstrate the possible importance of lipid peroxidation in activation of this compound (64).

7. AROMATIC AMINE METABOLISM

Studies on the metabolism of aromatic amines by PHS can be divided into two categories: those concerning compounds undergoing N-demethylation, such as secondary and tertiary aromatic amines, and those concerning compounds which undergo one electron oxidation, such as the carcinogenic primary aromatic amines.

We have reported that a wide variety of monomethyl- and dimethyl-substituted anilines, the drugs aminopyrine and benzphetamine, and the insecticide aminocarb are all N-demethylated by ram seminal vesicle microsomes fortified with arachidonic acid (34) (Table 1). N-demethylation was measured by production of formaldehyde. The reaction was dependent on the presence of intact microsomal protein and arachidonic acid or hydroperoxide; no other components were necessary in this system (34). The PHS inhibitors indomethacin and flufenamic acid drastically reduced formaldehyde production, while the cytochrome P-450 inhibitors metyrapone and SKF-525A were without effect (34). Using a variety of ring-substituted N-methylanilines it was observed that addition of substituents to the ring decreased metabolism of these compounds by PHS. Addition of a nitro group had the most dramatic effect; N-demethylation of N-methyl-o-nitroaniline and N-methyl-p-nitroaniline could not be detected (34). In a further investigation of the N-demethylation of p-chloro-N-methylaniline (PCMA), we demonstrated that very low concentrations of ram seminal vesicle microsomal protein were effective in carrying out this reaction. The amount of formaldehyde produced at a protein concentration of 0.15 mg/ml was 50% that at 2.0 mg/ml, the concentration at which maximal formaldehyde production was observed (34). PHS-dependent N-demethylation of PCMA was also observed using microsomal fractions from guinea pig and mouse lung, and rabbit kidney medulla (34). Although the rate of formaldehyde production by these tissues was only about 20% of that observed using RSVM, these tissues possess less than 1% of the PHS activity of RSVM. The fact that low concentrations of ram seminal vesicle microsomal protein and microsomal fractions from tissues with low PHS activity (i.e. compared to RSVM) both catalyze the N-demethylation of PCMA to a significant extent suggests a high turnover number for PHS peroxidase.

We have used the drug aminopyrine as a model compound with which to investigate the mechanism of PHS-catalyzed N-demethylations (50). As briefly described above, aminopyrine was first oxidized to the aminopyrine cation free radical (Figure 13). The radical was detected by electron spin

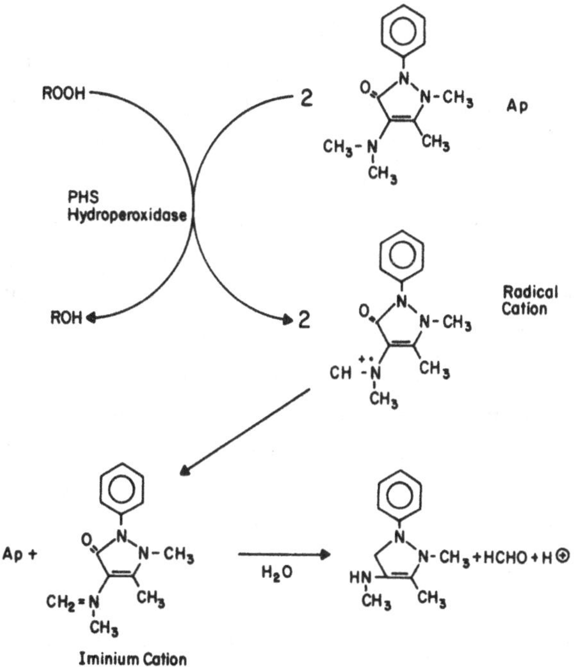

Figure 13. Mechanism for the N-demethylation of aminopyrene by PHS.

resonance and has strong absorbance at 565 nm. Thus, upon addition of arachidonic acid or peroxides to ram seminal vesicle microsomes, a blue color rapidly appears which then slowly fades. Two aminopyrene cation radicals can dispropor-tionate, yielding an iminium cation and aminopyrine. Alternatively, the aminopyrine radical can lose a second electron to form the iminium cation. This iminium cation is subsequently hydrolyzed to yield formaldehyde and the mono-methyl amine derivative. Thus water serves as the source of oxygen in the demethylation of aminopyrine. It is tempting to speculate that this mechanism applies to all the amines that undergo N-demethylation by PHS but further evidence is needed to support this conclusion.

Investigations on metabolism of primary aromatic amines by PHS have centered on two compounds: benzidine and 2-aminofluorene. Both of these compounds are potent carcinogens in animal studies (65), and epidemiological studies have demonstrated that the carcinogenic aromatic amines induce carcinomas of the urinary bladder in humans (66). We have suggested that PHS may serve as a pathway for the metabolic activation of these compounds to their ultimate carcinogenic forms in target tissues (44,67).

Josephy et al. have extensively investigated the metabolism of benzidine and its structural analogue 3,5,3´,5´-tetramethylbenzidine by horseradish peroxidase and PHS peroxidase (68,69). Both enzymes metabolize benzidine to a radical cation which is in equilibrium with a blue-colored charge transfer complex of benzidine itself and its two electron oxidation product, a diimine dication (Figure 14). The radical cations of both 3,5,3´,5´-tetramethylbenzidine (68) and benzidine (70) were observed directly by electron spin resonance spectroscopy. At less than equimolar peroxide, benzidine, the radical cation, the charge transfer complex and the diimine co-exist. At greater than equimolar peroxide, all the parent compound is oxidized to the 2-electron oxidation product, the diimine, which is yellow. Thus 3,5,3´5´-tetramethylbenzidine oxidation proceeds from blue to green to yellow. With benzidine, brown polymeric material was formed during oxidation. Azobenzidine was the only stable end product of peroxidation of benzidine identified (69) while the metabolites of 3,5,3´,5´-tetramethylbenzidine have not been identified. Benzidine oxidized by PHS in the presence of nucleic acids in vitro, yields reactive intermediates(s) that bind to the nucleic acid with very high efficiency (71,72). Benzidine or 3,5,3´,5´- tetramethylbenzidine oxidized by PHS

in the presence of certain phenol derivatives (i.e., butylated
hydroxyanisole, 2,6-dimethylphenol, serotonin, guaiacol, etc.)
formed adducts of an indoaniline type structure (Figure 14)

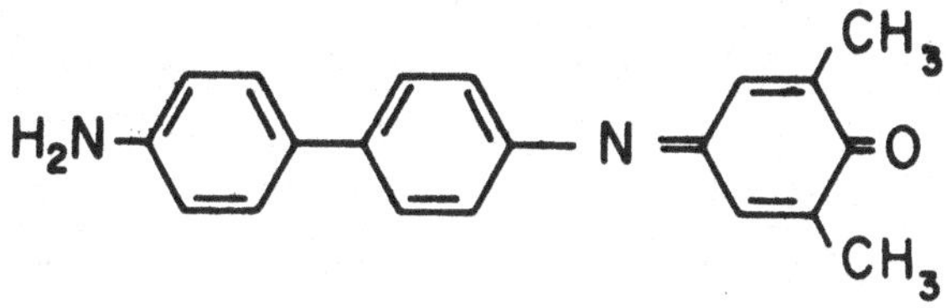

Figure 14.   Benzidine/butylated hydroxyanisole adduct.

(73,74). As discussed previously, PHS metabolized benzidine
to a mutagen as detected in a bacterial tester strain. It is
not currently known which of the intermediates pictured in
Figure 15 is responsible for mutagenicity, and adduct for-

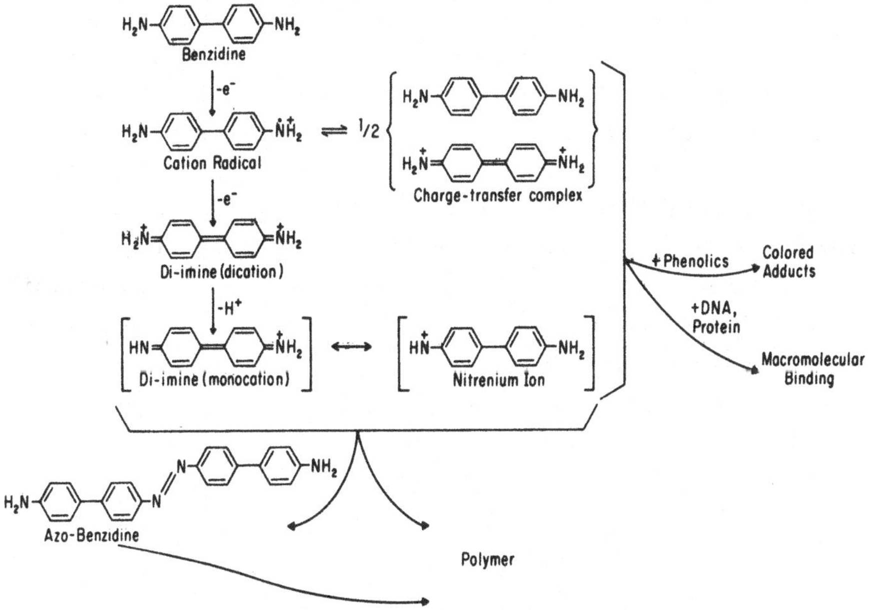

Figure 15.   Metabolism of benzidine by PHS and horseradish
             peroxidase.

mation with nucleic acids and phenols. Note that the loss of
a proton by the diimine dication results in formation of a
diimine monocation which is in equalibrium with a nitrenium
ion.

2-Aminofluorene is also metabolized by PHS. The major
isolable metabolites were 2,2´-azobisfluorene,
2-nitrofluorene; polymeric material and material covalently
bound to microsomal protein were also detected (44). The
reaction was dependent on arachidonic acid, was inhibited by
the addition of indomethacin to the incubation system, and was
very rapid. H_2O_2 also supported metabolism. These and other
data indicated that PHS peroxidase catalyzed the oxidation of
2-aminofluorene. N-hydroxy-2aminofluorene and
2-nitrosofluorene were extremely rapidly oxidized by PHS to
2-nitrofluorene. Horseradish peroxidase also oxidized
2-aminofluorene to 2-nitrofluorene and 2-2´-azobisfluorene,
but chloroperoxidase oxidized 2-aminofluorene to primarily
2-nitrosofluorene. This result indicates that PHS-dependent
oxidation of 2-aminofluorene may proceed through a free radical mechansim similar to that of horseradish peroxidase.

We have been unable to directly detect formation of a free
radical (most probably due to its extreme instability or
reactivity), but Boyd and Eling have generated several lines
of evidence suggesting one-electron oxidation is the mechansim
of metabolism of 2-aminofluorene by PHS (75). Metabolism of
2-aminofluorene by PHS also results in formation of intermediates which are mutagenic (48), bind covalently to nucleic
acids (71,72,76), and react with phenols to yield phenolic
adducts (75), (Figure 16).

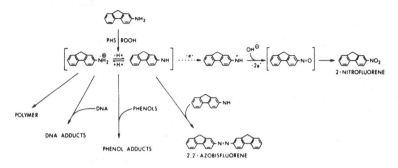

Figure 16. Metabolism of 2-aminofluorene by PHS and horseradish peroxidase.

Formation of N-hydroxy-2-aminofluorene could not be
detected during PHS-dependent oxidation of 2-aminofluorene
(44,75). Data derived from the formation of
2-aminofluorene/phenol adducts (75) and the identity of the
metabolites suggest that N-OH-2-AF is not formed to any significant extent, if at all, in this system. The reactive intermediates responsible for the binding to nucleic acids could be
free radicals or the 2-aminofluorene nitrenium ion.
Preliminary experiments have demonstrated that N-(deoxyguanosin-
8-yl)-2-aminofluorene (the sole DNA adduct formed by the
2-aminofluorene nitrenium ion) represents only a small fraction of the 2-aminofluorene/DNA adducts formed by PHS (76).
This suggests the nitrenium ion is not an important intermediate in the metabolism of 2-aminofluorene by PHS. Further
studies aimed at clarifying this point are underway.

8. METABOLISM OF PHENOLS

Phenols, like aromatic amines, are classical reducing
cofactors for peroxidases, including PHS. As seen in Table 1,
the metabolism of several phenolic compounds by PHS has been
studied. The analgesic acetaminophen is oxidized by PHS to a
reactive metabolite(s) that binds covalently to microsomal
protein (79), and reacts with reduced glutathione to give conjugates (80). PHS metabolizes acetaminophen to a phenoxyl
radical (12), which may be one of the reactive intermediates
involved. Other evidence suggests N-acetyl-p-

benzoquinonimine is an important reactive intermediate produced by oxidation of acetaminophen (11, and references
therein), and it is highly likely that the phenoxy radical
non-enzymatically oxidizes to form this compound.

Josephy et el. reported that p-aminophenol was oxidized by
PHS and horseradish peroxidase to the p-aminophenoxyl radical,
which then polymerizes (94). Indophenol was isolated in low
yields. Boyd and Eling (75) recently examined the metabolism
of the phenolic anti-oxidants butylated hydroxyanisole (BHA)
and 2,6-dimethylphenol. The major metabolites were dimers of
the parent compound, coupled at the position para to the phenolic group (See Table 1). In the case of BHA, this occurs
with loss of the methoxy group of one of the molecules of
parent compound.

Of particular interest is the PHS-mediated metabolism of
the transplacental carcinogen, diethylstilbestrol (DES). As
seen in Figure 5, DES, a diphenol, is oxidized via a semiquinone radical to the p-quinone. The p-quinone then
rearranges to β-dienestrol (39).

Peroxidative metabolism of DES, particularly by PHS, is
associated with its metabolic activation. We have studied the
metabolism of DES in Syrian Hamster Embryo (SHE) fibroblasts
(40), which are neoplastically transformed by DES (96). DES
was converted to β-dienestrol by these cells. The addition of
arachidonic acid enhanced formation of this metabolite.
Indomethacin inhibited metabolism of DES to trace levels, even
without addition of exogenous arachidonate, suggesting PHS was
responsible for basal levels of DES metabolism (40). Growing
SHE cells, which synthesize high levels of prostaglandins
(31), converted DES to β-dienestrol much more efficiently than
did confluent cultures, and indomethacin abolished metabolism
of DES in these cells as well (40). Other studies have
demonstrated that the ability of DES and related analogs to
transform SHE cells in culture does not correlate with the
reported estrogenic potency of these compounds in vivo, but
does correlate with the ability of these compounds to be metabolized by peroxidase-mediated oxidation (97).

Recently, Metzler and Epe investigated the peroxidase-mediated binding of DES analogs to DNA (98). This study suggested that the ability of these analogs to form quinones was not necessary for DNA binding, but that the ability to form a phenoxyl radical was required. Further work is needed to elucidate the ultimate reactive species of DES and its analogs, as well as the role of PHS in DES-induced carcinogenesis.

9. SUMMARY

It is clear from the work cited here and other chapters in this book that peroxidative metabolism and activation of toxins and carcinogens can occur in subcellular fractions and cell culture systems. Most of the work to date has been performed in these two systems. Although a role for peroxidase-mediated metabolism in vivo has not yet been established, detailed in vitro studies on mechanisms have yielded insights that allow the construction of experiments that are well designed to attack this problem. Future in vitro studies on the mechanism of cooxidation will continue to serve this role, as well as to further our understanding of the contribution of peroxidases in chemically induced toxicity and carcinogenicity.

TABLE 1. Chemicals Examined for Metabolism by PGH Synthase

Class	Chemical	Reaction	Product Isolated or Measured	Reference
Polyclyclic Aromatic Hydrocarbons	7,8-Dihydroxy-7,8-Dihydrobenzo [a]pyrene	Epoxidation	Anti-Diolepoxide	17,18
	7,8-Dihydrobenzo[a]pyrene	Epoxidation	9,10-Epoxy-7,8,9,10-tetrahydrobenzo-[a]pyrene	47
	1,2-Dihydroxy-1,2-Dihydro-chrysene	Epoxidation	Mutagen	45
	3,4-Dihydroxy-3,4-Dihydro-benzoanthracene	Epoxidation	Mutagen	45
	Benzo[a]pyrene	Hydroxylation	Quinones	41,42
	7,12-Dimethylbenzanthracene	Hydroxylation	7-Hydroxymethyl-12-methylbenzanthracenes	42
Sulfur containing compounds	Sulindac sulfide	Sulfoxidation	Sulindac	35
	Methylphenyl sulfide	Sulfoxidation	Methyl phenyl sulfoxide	36
	Bisulfite	Oxidation	Sulfur trioxide radical anion	77
	Lipoic acid	Oxidation	Thiosulfinate	78
Phenols	Phenidone	Dehydrogenation	1-Phenyl-3-hydroxy-pyrazole	38
	Epinephrine	Dehydrogenation	Adrenochrome	37
	Diethylstibestrol	Dehydrogenation	β-dienestrol	39
	Acetaminophen	Dehydrogenation	Covalent Binding, Glutathione conjugate	79,80
	Guaiacol	Dehydrogenation	3,3-Dimethoxydipheno-4,4´-quinone	20
	Butylated Hydroxylanisole (BHA)	Oxidation	2,2´-dihydroxy-5,5´-dimethoxy-3,3´-t-butylbiphenyl (Di-BHA)	81,75
	2,6-Dimethylphenol	Oxidation	3,5,3´,5´-tetramethyldipheno-4,4´-quinone	75
	p-Aminophenol	Oxidation	p-aminophenoxyl radical	94
	Phenetidine	Oxidation	GSH conjugate, dimer	82,83

Amines			
Benzidine	Oxidation	Diimine, radical cation	69
3,3,5,5'-Tetramethylbenzidine	Oxidation	Diimine, radical cation	68
2-aminofluorene	Oxidation	2-nitrofluorene; 2,2'-azobisfluorene	44
2-aminonaphthalene	Oxidation	2-nitrosonaphthalene, napthoquinonemine	71
4-aminobiphenyl	Oxidation	4-Nitrosobiphenyl	71
N-Methyl-4-aminoazobenzene	N-Demethylation	4-aminoazobenzene	71
N-Methylaniline	N-Demethylation	Formaldehyde	34
N-Methyl-p-chloroaniline	N-Demethylation	Formaldehyde	34
N-Methyl-o-nitroaniline	N-Demethylation	Formaldehyde	34
N-Methyl-p-nitroaniline	N-Demethylation	Formaldehyde	34
N-Methyl-o-toluidine	N-Demethylation	Formaldehyde	34
N-Methyl-m-toluidine	N-Demethylation	Formaldehyde	34
N-Methyl-p-toluidine	N-Demethylation	Formaldehyde	34
N-Methyl-o-phenylenediamine	N-Demethylation	Formaldehyde	34
N-Methyl-p-phenylenediamine	N-Demethylation	Formaldehyde	34
N-Methyl-m-nitrosoaniline	N-Demethylation	Formaldehyde	34
N,N-Dimethylaniline	N-Demethylation	Formaldehyde	34
N,N-Dimethylaminobenzoic acid	N-Demethylation	Formaldehyde	34
N,N-Dimethyl-p-nitroaniline	N-Demethylation	Formaldehyde	34
N,N-Dimethyl-p-phenylenediamine	N-Demethylation	Formaldehyde	34
Aminopyrine	N-Demethylation	Formaldehyde, cation free radical	34
Aminocarb (4-dimethylamino-m-tolyl-methylcarbamate)	N-Demethylation	Formaldehyde	34
Benzphetamine	N-Demethylation	Formaldehyde	34
Tryptophan	Oxidation	Radical cation	*

Misc.

	Reaction	Product	Ref
Quinacrine	Oxidation	Free radical	92
Tetramethylhydrazine	Demethylation	Free radical, formaldehyde	84
FANFT (N-[4-(5-nitro-2-furyl)-thiazolyl]formamide)	Oxidation	Unknown, DNA binding	85
ANFT (2-amino-4-(5-nitro-2-furyl) thiazole)	Oxidation	Unknown, DNA binding	85
Bilirubin	Oxidation	Unknown	86
Phenylbutazone	Oxidation	4-Hydroxy-phenylbutazone	43
Oxyphenylbutazone	Oxidation	4-Hydroxy-oxyphenylbutazone	87
Luminol	Dioxygenation	Aminophthalic acid	88
Diphenylisobenzofuran	Dioxygenation	Dibenzoylbenzene	88,89
Aflatoxin B_1	Epoxidation	DNA-adduct	90

*Eling TE, Mason RP: Unpublished observations

Table 2. Chemical Examined for Mutagenicity Using PHS as the Activiting Enzyme

Chemical	Relative Mutagenicity	Strain	Reference
Polycyclic Aromatic Hydrocarbons:			
Benzo[a]pyrene	−	TA98, TA100	45,46
Benzo[a]pyrene-7,8-diol	++++	TA98, TA100	45,46
Benzo[a]pyrene-4,5-diol	−	TA98	46,47
Benzo[a]pyrene-9,10-diol	−	TA98	4
Benzo[a]pyrene-7,8-dihydro	+++++	TA98	47
Benzo[a]pyrene-9,10-dihydro-	−	TA90	47
Benzo[a]anthracene	−	TA100	45
Benzo[a]anthracene-3,4-diol	+++	TA100	45
Benzo[a]anthracene-1,2-diol	−	TA100	45
Benzo[a]anthracene-8.9-diol	−	TA100	45
Benzo[a]anthracene-10,11-diol	−	TA100	45
Chrysene	−	TA100	45
Chrysene trans-1,2-diol	+++	TA100	45
Chrysene trans-3,4-diol	−	TA100	45
Chrysene trans-5,6-diol	−	TA100	45
Chrysene cis-5,6-diol	−	TA100	45
Aromatic Amines			
2-Aminofluorene	+++	TA98,TR-98**	48
2-Acetylaminofluorene	+	TA98	48
Benzidine	++	TA98	48
2,4,-Diaminoanisole	+	TA1538	48
2,5-Diaminoanisole	++	TA1538	48
1-Naphthylamine	−	TA1538	48
2-Naphthylamine	++	TA1538	48
Aniline	−	TA98	48
2-Aminoanthracene	−	TA98	48
Miscellanous			
Diethylstilbestrol	−	TA98	*
Phthalates	−	TA98	*
Dimethylnitrosamine	−	46	*

*Zeiger E, Eling TE: unpublished observations
**Boyd JA, Zeiger E, Eling TE: unpublished observations

REFERENCES

1. Doll R, Peto R: The causes of cancer: Quantitative estimates of avoidable risks of cancer in the United States today. J Nat'l Cancer Inst (66): 1193-1308, 1981.
2. Miller EC, Miller JA: Searches for ultimate chemical carcinogens and their reactions with cellular macromolecules. Cancer (47): 2327-2345, 1981.
3. Gelboin HV: Benzo(a)pyrene metabolism, activation, and carcinogenesis. Physiol Rev (60): 1107-1166, 1980.
4. Conney AH: Induction of microsomal enzymes by foreign chemicals and carcinogenesis by polycyclic aromatic hydrocarbons. Cancer Res (42): 4875-4917, 1982.
5. Phillips DH: Fifty years of benzo(a)pyrene. Natue (303): 468-472, 1983.
6. Weinstein IB, Jeffrey AM, Leffler S, Pulkrobek P, Yanasaki H, Grunberger D: Interactions between polycyclic aromatic hydrocarbons and cellular macromolecules. In: Gelboin HV, Ts'o POP (ed) Polycyclic hydrocarbons and cancer, Vol. 2. Academic Press, New York, 1978, pp. 3-36.
7. King CM: N-substituted aromatic compounds. In: Nicolini C (ed) Chemical carcinogenesis. Plenum Press, New York, 1982, pp. 25-46.
8. Beland FA, Beranek DT, Dooley KL, Heflich RH, Kadlubar FF: Arylamine-DNA adducts in vitro and in vivo: Their role in bacterial mutagenicity and urinary bladder carcinogenesis. Environ Health Perspec (49): 125-134, 1983.
9. Davidson DGD, Eastham WN: Acute liver necrosis following overdose of paracetamol. Br Med J (2): 497-499, 1966.
10. Mitchell JR, McMurtry RJ, Statham CN, Nelson SN: Molecular basis for several drug induced nephropathies. Am J Med (62): 518-526, 1972.
11. Dahlin DC, Miwa GT, Lu AYH, Nelson SD: N-acetyl-p-benzoquinone imine: A cytochrome P-450-mediated oxidation product of acetaminophen. Proc Natl Acad Sci USA (81): 1327-1331, 1984.
12. West PR, Harman LS, Josephy PD, Mason RP: Acetaminophen-enzymatic formation of a transient phenoxyl free radical. Mol Pharmacol, in press, 1984.
13. Boyd MR: Biochemical mechanisms in chemical-induced lung injury: Roles of metabolic activation. CRC Crit Rev Toxicol (7): 103-   , 1980.
14. Boyd MR: Biochemical mechanisms in pulmonary toxicity of furan derivatives. In: Hodgson E, Bend JR, Philpot RM (ed) Reviews in Biochemical Toxicology. Elsevier/North Holland, New York, 1980, pp. 71-102.
15. Sato R, Omura T: Cytochrome P-450. Academic Press, New York, 1978.
16. Van der Ouderaa FJ, Buytenhek M, Nugteren DH, Van Dorp DA: Purification and characterization of prostaglandin endoperoxide synthetase from sheep vesicular glands. Biochim Biophys Acta (487): 315-331, 1977.
17. Marnett LJ, Johnson JT, Bienkowski MJ: Arachidonic acid-dependent metabolism of 7,8-dihydroxy-7,8-

dihydrobenzo(a)pyrene by ram seminal vesicles. FEBS Letts (106): 13-16, 1979.

18. Sivarajah K, Mukhtar H, Eling T: Arachidonic acid-dependent metabolism (±)-trans-7,3-dihydroxy-7,8-dihydro-benzo(a)pyrene (BP-7,8-diol) to 7,10/8,9 tetrols. FEBS Letts (106): 17-20, 1979.

19. Christ EJ, Van Dorp DA: Comparative aspects of prostaglandin biosynthesis in animal tissue. Biochim Biophys Acta (270): 537-541, 1972.

20. Ohki S, Nobuchika O, Yamamoto S, Hayaishi O: Prostaglandin hydroperoxide, an integral part of prostaglandin endoperoxide synthetase from bovine vesicular gland microsomes. J Biol Chem (254): 829-836, 1979.

21. Kulmarz RJ, Lands WEM: Characteristics of prostaglandin H synthase. In: Samuelsson B, Paoletti R, Ramwell P (ed) Advances in prostaglandin, thromboxane and leukotriene research, Vol. 11. Raven Press, New York, 1983, pp. 93-97.

22. Samuelsson B: Leukotrienes: Mediators of immediate hypersensitivity reactions and inflammation. Science (220): 568-575, 1983.

23. Bryant RW, Simon TC, Bailey JM: Role of glutathione peroxidase and hexose monophosphate shunt in the platelet lipoxygenase pathway. J Biol Chem (257): 14937-14943, 1982.

24. Bryant RW, Simon TC, Bailey JM: Hydroperoxy fatty acid formation in selenium deficient rat platelets: Coupling of glutathione peroxidase to the lipoxygenase pathway. Biochem Biophys Res Commun (117): 183-189, 1983.

25. Mizuno K, Yamamoto S, Lands WEM: Effects of non-steroidal antiinflammatory drugs on fatty acid cyclooxygenase and prostaglandin hydroperoxidase activities. Prostaglandins (23): 743-757, 1983.

26. Hope WC, Welton AF, Fiedler-Nagy C, Batula-Bernarbo C, Coffey J: In vitro inhibition of the biosynthesis of slow reacting substance of anaphylaxis (SRS-A) and lipoxygenase activity by quercetin, Biochem Pharmacol (32): 367-371, 1983.

27. Hong FL, Carty T, Degykin D: Tranylcypramine and 15-hydroperoxy-arachidonic acid affect arachidonic acid release in addition to inhibition of prostacyclin synthetase in calf aortic endothelial cells. J Biol Chem (255): 4538-4540, 1980.

28. Bills TK, Smith JB, Silver JB: Metabolism of [^{14}C] arachidonic acid by human platelets. Biochim Biophys Acta (424): 303-314, 1976.

29. Levine L: Arachidonic acid transformation and tumor production. Adv Cancer Res (35): 49-79, 1981.

30. Piper PJ, Vane JR: The release of prostaglandins from lung and other tissues. Ann NY Acad Sci (180): 363-385, 1971.

31. Korbut R, Boyd J, Eling T: Respiratory movements alter the generation of prostacyclin and thromboxane A_2 in isolated rat lungs: The influence of arachidonic acid-pathway inhibitors on the ratio between pulmonary prosta-

cyclin and thromboxane A_2. Prostaglandins (21): 491-503, 1981.

32. Ali AE, Barrett JC, Eling TE: Prostaglandin and thromboxane production by fibroblasts and vascular endothelial cells. Prstaglandins (20): 667-688, 1980.

33. Hong SL, Levine L: Inhibition of arachidonic acid release from cells as the biochemical action of anti-inflammatory corticosteroids. Proc Natl Acad Sci USA (73): 1730-1734, 1976.

34. Sivarajah K, Lasker JM, Eling TE, Abou-Donia MB: Metabolism of N-alkyl compounds during the biosynthesis of prostaglandins. Mol Pharmacol (21): 133-141, 1982.

35. Egan RW, Gale PH, VandenHeuval WJA, Baptista EM, Kuehl Jr. FA: Mechanism of oxygen transfer by prostaglandin hydroperoxidase. J Biol Chem (255): 323-326, 1980.

36. Egan RW, Gale PH, Baptista EM, Kennicott KL, VandenHeuval WJA, Walter RW, Fagerness PE, Kuehl, Jr FA: Oxidation reactions by prostaglandin cyclooxygenase-hydroperoxidase. J Biol Chem (256): 7352-7361, 1981.

37. Porter NA, Wolf RA, Pagels WR, Marnett LJ: A test for the intermediacy of 11-hydroperoxyeicosa-5,8,12,14-tetraenoic acid (11-HPETE) in prostaglandin biosynthesis. Biochem Biophys Res Commun (92): 349-355, 1980.

38. Marnett LJ, Siedlik PH, Fung LWM: Oxidation of phenidone and BW755C by prostaglandin endoperoxide synthetase. J Biol Chem (257): 6957-6964, 1982.

39. Degen GH, Eling TE, McLachlan JA: Oxidative metabolism of diethylstilbestrol by prostaglandin synthetase. Cancer Res (42): 919-923, 1982.

40. McLachlan JA, Wong A, Degen GH, Barrett JC: Morphological and neoplastic transformatin of Syrian hamster embryo fibroblasts by diethylstilbestrol and its analogues. Cancer Res (42): 3040-3045, 1982.

41. Marnett LJ, Reed GA, Johnson JT: Prostaglandin synthetase dependent benzo(a)pyrene oxidation: Products of the oxidation and inhibition of their metabolism by antioxidants. Biochem Biophy Res Commun (79): 569-576, 1977.

42. Sivarajah K, Anderson MW, Eling TE: Metabolism of benzo(a)pyrene to reactive intermediate(s) via prostaglandin biosynthesis. Life Sci (23): 2571-2578, 1978.

43. Marnett LJ, Bienkowski MJ, Pagels WR, Reed, GA: Mechanism of xenobiotic cooxidation coupled to prostaglandin H_2 biosynthesis. In: Samuelsson B, Ramwell P, Paoletti R (ed) Advances in prostaglandin and thromboxane research, Vol. 6. Raven Press, New York, 1980, pp. 149-151.

44. Boyd JA, Harvan DJ, Eling TE: The oxidation of 2-aminofluorene by prostaglandin endoperoxide synthetase. J Biol Chem (258): 8246-8254, 1983.

45. Guthrie J, Robertson IGC, Zeiger E, Boyd JA, Eling TE: Selective activation of some dihydrodiols of several polycyclic aromatic hydrocarbons to mutagenic products by prostaglandin synthetase. Cancer Res (42): 1620-1623, 1982.

46. Marnett LJ, Reed GA, Dennison DJ: Prostaglandin synthe-

tase dependent activation of 7,8-dihydro-7,8-dihydroxy-benzo(a)pyrene to mutagenic derivatives. Biochem Biophys Res Commun (82): 210-216, 1978.

47. Reed GA, Marnett LJ: Metabolism and activation of 7,8-dihydrobenzo(a)pyrene during prostaglandin biosynthesis. J Biol Chem (257): 11368-11376, 1982.

48. Robertson IGC, Sivarajah K, Eling TE, Zeiger E: Activation of some aromatic amines to mutagenic products by prostaglandin endoperoxide synthetase. Cancer Res (43): 476-480, 1983.

49. Rahimtula A, Moldeus P, Andersson B, Nordenskjold M: Prostaglandin synthetase catalyzed DNA strand breaks by aromatic amines. In: Powles T, Beckman R, Honn K, Ramwell P (eds) Prostaglandins and cancer: The first international conference. Alan R, Liss, Inc., New York, 1982, pp. 159-162.

50. Lasker JM, Sivarajah K, Mason RP, Kalyanaraman B, Abou-Donia MB, Eling TE: A free radical mechanism of prosta-glandin synthetase-dependent aminopyrine demethylation. J Biol Chem (256): 7764-7767, 1981.

51. Dix TA, Marnett LJ: Free radical epoxidation of 7,8-dihydroxy-7,8-dihydrobenzo(a)pyrene by hematin and polyunsaturated fatty acid hydroperoxides. J Amer Chem Soc (103): 6744-6746, 1981.

52. Reed GA, Brooks EA, Eling TE: Phenylbutazone-dependent epoxidation of 7,8-dihydroxy-7,8-dihydrobenzo(a)pyrene. A new mechanism for prostaglandin H synthase-catalyzed oxidations. J Biol Chem (259): 5591-5595, 1984.

53. Sivarajah K, Lasker JM, Eling TE: Prostaglandin synthetase-dependent cooxidation of (±)-benzo(a)pyrene-7,8-dihydrodiol by human lung and other mammalian tissues. Cancer Res (41): 1834-1839, 1981.

54. Boyd JA, Barrett JC, Eling TE: Prostaglandin endoper-oxide synthetase dependent cooxidation of (±)-trans-7,8-dihydroxy-7,8-dihydrobenzo(a)pyrene in C3H 10T$\frac{1}{2}$ clone 8 cells. Cancer Res (42): 2628-2632, 1982.

55. Sivarajah K, Jones KG, Fouts JR, Deveroux T, Shirley JE, Eling TE: Prostaglandin synthetase and cytochrome P-450-dependent metabolism of (±)-benzo(a)pyrene-7,8-dihydrodiol by enriched populations of rat clara cells and alveolar type II cells. Cancer Res (43): 2632-2636, 1983.

56. Reed GA, Grafstrom RC, Krauss RS, Autrup H, Eling TE: Prostaglandin H synthase-dependent co-oxygenation of (±)-7,8-dihydroxy-7,8-dihydrobenzo(a)pyrene in hamster trachea and human bronchus explants. Carcinogenesis (5): 955-960, 1984.

57. Adriaenssens PI, Sivarajah K, Boorman GA, Eling TE, Anderson MW: Effect of aspirin and indomethacin on the formation of benzo(a)pyrene-induced pulmonary adenomas and DNA adducts in A/HeJ mice. Cancer Res (43): 4762-4767, 1983.

58. Wiley MH, Feingold KR, Grunfeld C, Quesney-Huneeus V, Wu JM: Evidence for cAMP-independent inhibition of S-phase

DNA synthesis by prostaglandins. J Biol Chem (258): 491-496, 1983.

59. Andreis PG, Whitfield JF, Armato U: Stimulation of DNA synthesis and mitosis of hepatocytes in primary cultures of neonatal rat liver by arachidonic acid and prostaglandins. Exp Cell Res (134): 265-272, 1981.

60. De Asua CJ, Richmond KMV, Otto AM: Two growth factors and two hormones regulate initiation of DNA synthesis in cultured mouse cells through different pathways of events. Proc Natl Acad Sci USA (78): 1004-1008, 1981.

61. Otto AM, Nilsen-Hamilton M, Boss BD, Ulrich MO, De Asua LJ: Prostaglandins E_1 and E_2 interact with prostaglandin $F_{2\alpha}$ to regulate initiation of DNA replication and cell division in Swiss 3T3 cells. Proc Natl Acad Sci USA (79): 4992-4996, 1982.

62. Fischer SM, Gleason GC, Hardin LG, Bohrman JS, Slaga TJ: Prostaglandin modulation of phorbol ester skin tumor promotion. Carcinogenesis (1): 245-248, 1980.

63. Fischer SM, Mills GD, Slaga TJ: Inhibition of mouse skin tumor promotion by several inhibitors of arachidonic acid metabolism. Carcinogenesis (3): 1243-1245, 1982.

64. Dix TA, Marnett LJ: Metabolism of polycyclic aromatic hydrocarbon derivatives to ultimate carcinogens during lipid peroxidation. Science (221): 77-79, 1983.

65. Clayson DB, Garner RC: Carcinogenic aromatic amines and related compounds. In: Searle CE (ed) Chemical carcinogens, ACS Monograph 173, American Chemical Society, Washington DC 1976, pp. 366-461.

66. Radomski JL: The primary aromatic amines: Their biological properties and structure-activity relationships. Ann Rev Pharmacol Toxicol (19): 129-157, 1979.

67. Krauss RS, Eling TE: Arachidonic acid-dependent cooxidation: A potential pathway for the activation of chemical carcinogens in vivo. Biochem Pharmacol, in press, 1984.

68. Josephy PD, Mason RP, Eling TE: Cooxidation of the clinical reagent 3,5,3',5'-tetramethylbenzidine by prostaglandin synthase. Cancer Res (42): 2567-2570, 1982.

69. Josephy PD, Eling TE, Mason RP: Cooxidation of benzidine by prostaglandin synthase and comparison with the action of horseradish peroxidase. J Biol Chem (258): 5561-5569, 1983.

70. Wise RW, Zenser TV, Davis BB: Prostaglandin H synthase metabolism of the urinary bladder carcinogens benzidine and ANFT. Carcinogenesis (4): 285-289, 1983.

71. Kadlubar FF, Frederick CB, Weiss CC, Zenser TV: Prostaglandin endoperoxide synthetase-mediated metabolism of carcinogenic aromatic amines and their binding to DNA and protein. Biochem Biophys Res Commun (108): 253-258, 1982.

72. Morton KC, King CM, Vaught JB, Wang CY, Lee MS, Marnett LJ: Prostaglandin H synthase-mediated reaction of carcinogenic arylamines with tRNA and homopolyribonucleotides. Biochem Biophys Res Commun (111): 96-103, 1983.

73. Josephy PD, Mason RP, Eling TE: chemical structure of

the adducts formed by the oxidation of benzidine in the presence of phenols. Carcinogenesis (3): 1227-1230, 1982.

74. Josephy PD, Eling TE, Mason RP: An electron spin resonance study of the activation of benzidine by peroxidases. Mol Pharmacol (23): 766-770, 1983.

75. Boyd JA, Eling TE: Evidence for a one-electron mechanism of 2-aminofluorene oxidation by prostaglandin H synthase and horseradish peroxidase. J Biol Chem, in press, 1984.

76. Krauss RS, Reed GA, Eling TE: Formation of unique arylamine/DNA adducts from 2-aminofluorene activated by prostaglandin H synthase. Proc Am Assoc Cancer Res (25): 84, 1984.

77. Mottley C, Mason RP, Chignell CF, Sivarajah K, Eling TE: The formation of sulfur trioxide radical anion during the prostaglandin hydroperoxidase-catalyzed oxidation of bisulfite (hydrated sulfur dioxide). J Biol Chem (257): 5050-5055, 1982.

78. Marnett LJ, Wilcox CL: Stimulation of prostaglandin biosynthesis by lipoic acid. Biochim Biophys Acta (478): 222-230, 1977.

79. Boyd, JA, Eling TE: Prostaglandin endoperoxide synthetase-dependent cooxidation of acetaminophen to intermediates which covalently bind in vitro to rabbit renal medullary microsomes. J Phamacol Exp Ther (219): 659-664, 1981.

80. Moldeus P, Rahimtula A: Metabolism of paracetamol to a glutathione conjugate catalyzed by prostaglandin synthetase. Biochem Biophys Res Commun (96): 659-664, 1980.

81. Rahimtula A: In vitro metabolism of 3-t-butyl-4-hydroxyanisole and its irreversible binding to proteins. Chem-Biol Inter (45): 125-135, 1983.

82. Andersson B, Larsson R, Rahimtula A, Moldeus P: Hydroproxide-dependent activation of p-phenetidine catalyzed by prostaglandin synthase and other peroxidases. Biochem Pharmacol (32): 1045-1050, 1983.

83. Andersson B, Carsson R, Rahimtula A, Moldeus P: Prostaglandin synthase and horseradish peroxidase catalyzed DNA-binding of p-phenetidine. Carcinogenesis (5): 161-165, 1984.

84. Kalyanaraman B, Sivarajah K, Eling TE, Mason RP: A free radical mediated cooxidation of tetramethylhydrazine by prostaglandin hydroperoxidase. Carcinogenesis (4): 1341-1343, 1983.

85. Zenser TV, Paler MO, Mattammal MB, Bolla RI, Davis BB: Comparative effects of prostaglandin H synthase-catalzed binding of two 5-nitrofuran urinary bladder carcinogens. J Pharmacol Exp Ther (227): 139-143, 1983.

86. Reed GA, Lasker JM, Eling TE, Sivarajah K: Peroxidative oxidation of bilirubin: A spectrophotometric assay for prostaglandin H synthase. Prostaglandins, in press, 1984.

87. Portoghese PS, Svanborg K, Samuelsson B: Oxidation of oxyphenbutazone by sheep vesicular gland microsomes and lipoxygenase. Biochem Biophys Res Commun (63): 748-755, 1975.

88. Marnett LJ, Wlodawer P, Samuelsson B: Co-oxygenation of

organic substrates by the prostaglandin synthetase of sheep vesicular gland. J Biol Chem (250): 8510-8517, 1975.

89. Marnett LJ, Bienkowski MJ, Pagels WR: Oxygen 18 investigation of the prostaglandin synthetase-dependent co--oxidation of diphenylisobenzofuran. J Biol Chem (254): 5077-5082, 1979.

90. Amstad P, Cerutti P: DNA binding of aflatoxin B_1 by co-oxygenation in mouse embryo fibroblasts $C3H/10T^1/_2$. Biochem Biophys Res Commun (112): 1034-1040, 1983.

91. Marnett LJ, Bienkowski MJ: Hydroperoxide dependent oxygenation of trans-7,8-dihydroxy-7,8-dihydrobenzo(a)pyrene by ram seminal vesicle microsomes. Source of the oxygen. Biochem Biophy Res Commun (96): 639-697, 1980.

92. Sinha BK, Irreversible binding of quinacrine to nucleic acids during horseradish peroxidase- and prostaglandin synthetase-catalyzed oxidation. Biochem Pharmacol (32): 2604-2607, 1983.

93. Marnett LJ, Reed GA: Peroxidatic oxidation of benzo(a)-pyrene and prostaglandin biosynthesis. Biochemistry (18): 2923-2929, 1979.

94. Josephy PD, Eling TE, Mason RP: Oxidation of p-aminophenol catalyzed by horseradish peroxidase and prostaglandin synthase. Mol Pharmacol (23): 461-466, 1983.

95. Panthananickal A, Marnett LJ: Arachidonic acid-dependent metabolism of ($\pm$)-7,8-dihydroxy-7,8-dihydrobenzo(a)pyrene to polyguanylic acid-binding derivatives. Chem Biol Inter (33): 239-252, 1981.

96. Barrett JC, Wong A, McLachlan JA: Diethylstilbestrol induces neoplastic transformation without measurable gene mutation at two loci. Science (212): 1402-1404, 1981.

97. McLachlan JA, Wong A, Degen GH, Barrett JC: Morphological and neoplastic transformation of syrian hamster embryo fibroblasts by diethylstilbestrol and its analogs. Cancer Res (42): 3040-3045, 1982.

98. Metzler, M, Epe B: Peroxidase-mediated binding of diethyl-stilbestrol analogs to DNA in vitro: A possible role for a phenoxyl radical. Chem Biol Inter (50): 351-360, 1984.

4

PROSTAGLANDIN SYNTHASE-DEPENDENT COOXIDATION AND AROMATIC AMINE
CARCINOGENESIS

JOHN R. RICE, TERRY V. ZENSER AND BERNARD B. DAVIS

1. INTRODUCTION AND OVERVIEW
1.1. Aromatic amines and bladder cancer

Aromatic amines were among the first chemical agents
recognized to be carcinogenic. This suspicion was first voiced
by Rehn in 1895 (1) after he noticed three cases of bladder
cancer among workers employed at an amine-derived dye factory in
Germany. Prior to the industrialized production of synthetic
organic chemicals (thus occasioning long-term exposure of significant amounts to many people), this disease was sufficiently
rare to make the situation noteworthy. As large-scale human
exposure continued, particularly in the virtual absence of sound
hygenic practices, enough cases of malignant disease arose to
permit an epidemiological assignment of cause. In this way,
4-aminobiphenyl was identified as a human carcinogen (and its
use banned) in 1955 (2), and both benzidine and 2-naphthylamine
were similarly classified, following indications derived from
animal models, in 1954 (3). The dog was initially selected as a
test species of choice for chemically-induced bladder cancer
because of its sensitivity to the disease. Dosing studies of
the three compounds cited above showed that 4-aminobiphenyl is
the most potent, followed by 2-naphthylamine (4), while benzidine
clearly shows a lower order of response (5). These studies, and
those of the induction of liver tumors in rodents (6), indicate
that individual aromatic amines of superficially similar structure differ widely in their carcinogenic potential. In addition,
the disparate sites of tumor formation shown by these compounds
indicates that they require enzymatic transformation prior to
eliciting their effects, as described in more detail in other
chapters in this volume.

Recognition of the need for metabolic transformation has led
to biochemical studies of the specific enzymes that appear to be
important in this process. The previous decade has witnessed an
accumulation of data indicating that numerous carcinogens and
other xenobiotic compounds may be cooxidatively metabolized by
the hydroperoxidase activity of prostaglandin synthase
(previously called prostaglandin cyclooxygenase or prostaglandin
endoperoxide synthetase). This chapter will briefly review the
tissue distribution of this enzyme and the bases for the
hypothesis that the activity of this enzyme is involved in the
initiation phase of aromatic amine-induced bladder cancer.
Particular emphasis will be placed on the molecular events which
have been elucidated to occur during this process or which
appear, by analogy with known chemical principles or the
mechanism of reactions mediated by other hemoprotein enzymes, to
explain the existing data. It is intended that this perspective
will complement the other chapters of this volume as well as
several excellent reviews that have recently appeared,
particularly those of Marnett (7), Gale and Egan (8), and
Marnett and Eling (9).

1.2. Distribution and characteristics of prostaglandin synthase

Prostaglandin synthase is widely distributed in animal, plant,
and bacterial cells, with the highest concentrations found in the
mammalian seminal vesicle (or vesicular gland), platelet, and
renal medulla, and lesser amounts in brain, spleen, and lung
(10). This activity has also been shown to occur in the bladder
transitional epithelium of dogs (11), rats and rabbits (12), and
hamsters and guinea pigs (T.V. Zenser, unpublished observations).
The native enzyme is a membrane-bound glycoprotein which, when
solubilized with detergent, exists as a dimer of two 70,000-Da
subunits (13). The fully functional enzyme requires as a cofactor one molecule of heme per dimer (14). It has also been
located on the nuclear envelope (15), and thus may occur within
the nucleus (7) in close proximity to nuclear DNA. The synthase
complex catalyzes the oxygenation of polyunsaturated fatty acids
which have in common a triene system terminating at the sixth

carbon from the alkyl terminus. Quantitatively, the most important of these is arachidonic acid, which is released from the phospholipid component of the plasma membrane by lipases. This well-known reaction occurs in two steps, each mediated by the synthase, as depicted in Figure 1. The first step, termed the cyclooxygenase activity, results in the addition of two molecules of molecular oxygen to the fatty acid chain to yield prostaglandin (PG) G_2. This phase of the reaction is specifically inhibited by several agents, including aspirin and indomethacin. The molecular mechanism of the irreversible inhibition by aspirin involves the transfer of an acetyl group to a serine residue of the enzyme polypeptide chain (16). In the second step, the enzyme appears to function as a classical hydroperoxidase, with general specificity for a broad range of peroxide-containing substrates. With PGG_2, this results in the reductive cleavage of the hydroperoxide function to the alcohol (PGH_2). This product may then be further modified by a variety of enzymes that participate in the "arachidonic acid cascade" to ultimately yield a variety of prostaglandins and thromboxanes (8). The hydroperoxidase step requires the availability of two reducing equivalents per cleavage reaction, which may be acquired via the alteration of a cosubstrate molecule. This metabolic conversion of the cosubstrate is depicted in Figure 1 as the conversion of A to B. A suitable cosubstrate may provide the reducing equivalents by releasing one or two electrons, or it may combine with the reactive oxygen-containing entity that is cleaved from the peroxide. In this way, as the fatty acid substrate is oxidized by the addition of molecular oxygen, the cosubstrate becomes "cooxidized" by either oxygen insertion or electron withdrawal. Several specific examples of exogenous cosubstrate cooxidation by each of these mechanisms have been reported, although the endogenous reducing cofactor(s) that serve in the absence of xenobiotic organic molecules have not been identified. In many cases, the altered cofactor product B is a highly reactive electrophilic intermediate capable of very rapid reaction with cellular constituents (including the synthase itself), or it may be relatively unreactive and hence amenable to isolation and characterization. These metabolites could potentially be involved in many reactions that are deleterious to cells, and as such could be mediators of the influence that prostaglandin synthase has upon the processes of tissue necrosis, aging and carcinogenesis.

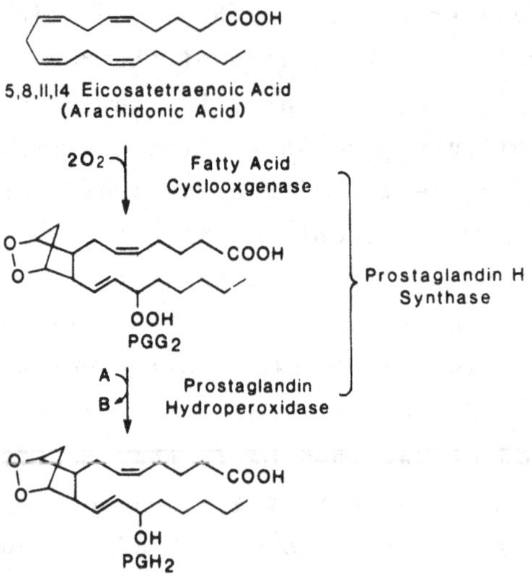

FIGURE 1. Conversion of organic cosubstrate "A" to oxidized product "B" by the hydroperoxidase component of prostaglandin synthase.

A pertinent consideration in understanding the involvement of prostaglandin synthase in the metabolism of aromatic amines is the mechanistic cycle of the hydroperoxidase activity. This enzyme appears to be mechanistically related to a number of other, more extensively studied hemoprotein enzymes such as horseradish peroxidase, lactoperoxidase, cytochrome c peroxidase, and others. Many of these are much easier to isolate and study and thus have been utilized as models of the hydroperoxidase mechanism. Prostaglandin synthase is unique within this group, being both membrane-bound in its native state and able to generate its own peroxide substrate (via its cyclooxygenase component) from suitable fatty acids. An excellent review of

the mechanism of these enzymes has been provided by Walsh (17).
Despite subtle differences in peroxide substrate specificity
that may be due to dissimilarities in their polypeptide chains,
all possess a ferriheme prosthetic group and catalyze the
reductive cleavage of a wide variety of alkyl peroxides to the
coresponding alcohol. Hydrogen peroxide (H_2O_2) is also a
substrate and is reduced to water. The most complete picture of
this process has been obtained with horseradish peroxidase. In
its resting state, the heme-bound iron atom of the enzyme is in
the +3 state. Following binding of the peroxide to the active
site, the O-O bond is cleaved, and a water molecule is released
with concomitant oxidation of the ferriheme moiety by two
equivalents. The resulting spectroscopically-distinct complex
is called Compound I. The exact electronic structure of this
highly-oxidized transient state sparked controversy for many
years, but application of a variety of physical methods have
resulted in general agreement that one electron is lost from the
iron atom and one from the porphyrin ring, the latter yielding a
cation radical observable with electron spin resonance spectroscopy (18). A corresponding, although not identical, transient
state has also been observed with prostaglandin synthase (19).
In the presence of an organic substrate capable of releasing a
single electron, Compound I is reduced to Compound II, which is
thus one-electron-oxidized relative to the resting state.
Aquisition of a second electron from the same or a different
cosubstrate molecule reduces the heme iron back to the +3 state
and thus completes the enzymatic cycle.

Several important aspects of the generalized hydroperoxidase
mechanism remain obscure, and one complication is that most of
these probably depend upon the particular enzyme involved and
upon the individual cosubstrate molecule under consideration.
One question concerns the nature of the peroxide-derived oxy
radical that is associated with the Compound I intermediate
(20), the way in which this species is bonded to the enzyme-
cosubstrate complex, and the manner in which this component is
released as a product. As described in the following section,
with some cosubstrates an oxygen atom is ultimately inserted
into the cosubstrate molecule, while other cosubstrates prompt

the incorporation of this atom into water.

A second question concerns the similarities that may exist between the mechanistic cycle of the peroxidases and that of the well-known mixed-function oxidase, cytochrome P-450. In addition to its ability to hydroxylate substrates by a mechanism requiring molecular oxygen and NADPH, this cytochrome also exhibits peroxidase activity (20). Certain aspects of the P-450 peroxidase mechanism are clearly different from those of the classical hydroperoxidases, notably the lack of long-lived and spectrally-distinct intermediates corresponding to Compounds I and II during the metabolism of some cosubstrates (20,21). However, it is clear that both the peroxide-dependent and O_2-dependent P-450 mechanisms include a stage in which a reactive oxygen species is incorporated into suitable cosubstrates. This intermediate is also highly electron deficient, and thus may also be capable of withdrawing electrons from certain other organic cosubstrates. Therefore, it appears reasonable that both P-450 and some peroxidases may create the same metabolic products from the same cosubstrates, despite subtle differences in the arrangements of atoms and electrons in the vicinity of each enzyme's active site. The overall consequence may be hydroxylation via oxygen insertion, or release of electrons by the cosubstrate, such that mechanistic information about product formation derived from one system may be a reasonable inference about a possible corresponding mechanism of the other. The following section will review examples of prostaglandin synthase-catalyzed metabolism by both oxygen insertion and/or electron withdrawal. In several cases, the peroxidase data closely parallels the P-450 data.

2. IN VITRO STUDIES OF PROSTAGLANDIN SYNTHASE-DEPENDENT METABOLISM

The overwhelming majority of studies on the function of prostaglandin synthase as a xenobiotic-metabolizing enzyme system have utilized broken-cell preparations, although tissue slice experiments have also been reported (22). The location of the enzyme on the endoplasmic reticulum has made use of the microsomal fraction a common means of partially extricating the active material from extraneous cellular constituents. Other studies have successfully employed the purified enzyme (23). For practical reasons, many investigators employ the ram seminal vesicle as a particularly rich source of synthase activity. Others have prepared the enzyme from the inner medulla of rabbit kidneys (24-26).

A principal reason for the popularity of in vitro incubations is the flexibility with which the investigator may control the relative amounts of the constituents. Common incubations contain more or less arbitrary amounts of buffer salt, enzyme, fatty acid or hydroperoxide substrate, xenobiotic cosubstrate, a source of heme, and sometimes an inhibitor or an endogenously-occurring nucleophilic trap (e.g., glutathione). Depending on the means by which the investigator is examining the system, he may optimize the amounts of some constituents so as to maximize the yield of or sensitivity to the experimental "observable". The latter commonly include the consumption of molecular oxygen (via an O_2-sensitive electrode probe), quantitation of the yield of prostaglandin products, production of extractable metabolites of an organic cosubstrate, appearance of a free radical magnetic resonance signal, or covalent reaction of radiolabelled material with cellular constituents. In fact, it is usually possible to adjust conditions such that any predetermined amount of conversion of the organic cosubstrate, from 0 to 100%, may be achieved. This is a major problem, because it is impossible to know the degree to which such "optimized" conditions reflect the realistic situation within a functioning whole-cell system or those that could actually exist in vivo. An additional problem is that investigations frequently are focussed on only one or two of the above experimental observables, despite the fact that several of the others may also be occurring, and which may be of even greater significance. In spite of these potential pitfalls and the gaps in under-standing which occasionally result, most of what is known about prostaglandin synthase has been derived from in vitro studies.

2.1. Cooxidation by oxygen insertion

Metabolic oxidation is an extensively-studied route of biotransformation. The goal of this process is to render lipophilic, non-polar molecules of both endogenous and exogenous origin more polar and water soluble, thus facilitating their excretion by the kidneys. For many molecules, a means to this end is the enzymatic addition of a hydroxy group. In many cases, the latter is further conjugated with sulfate or glucuronide as a means of imparting more polar character to the substrate. Many such hydroxylation steps occur in the liver and are mediated by the cytochrome P-450 system. By contrast, far fewer substrates have been shown to be hydroxylated by peroxidases. However, due to the relative stability of certain oxygenated metabolites, several early studies of the biotransformation of organic cosubstrates by prostaglandin synthase successfully isolated such products. There appear to be at least two mechanisms by which the enzyme mediates the incorporation of oxygen into a substrate molecule. One of these occurs with a suitable substrate bound to the active site. In the other, the oxygen insertion reaction occurs by a radical-propagating process in free solution following the release of an altered substrate molecule by the enzyme. It is important to recognize the distinction between these mechanisms and that which apparently operates with aromatic amine substrates (next section).

Two cosubstrates which are oxygenated while bound to the enzyme are sulindac sulfide (27,28) and methyl phenyl sulfide (23). This process has been extensively characterized in studies by Egan and coworkers (8 and references therein). With each of these substrates, experiments performed with an analogue of PGG_2 specifically labelled with $^{18}O_2$ have conclusively shown that the oxygen atom that was added to the sulfide originated exclusively from the oxygen of the hydroperoxide group (as opposed to originating from either solvent H_2O or dissolved O_2). An identical 1:1 correspondence between the amounts of the sulfoxide and fatty acid products that were formed indicates that the sulfide cosubstrate accounted for all of the reducing equivalents required by the enzyme for the peroxide reduction (27). Apparently, organic sulfides are resistant to the release of electrons and thus provide the required reducing equivalents by reacting directly with an oxygen-containing entity that is bound to the heme iron

following the homolytic cleavage of the peroxide group. This
process is facilitated by an associative binding of the sulfide
to the enzyme, which thereby expedites transfer of the oxidant
to the sulfide and accounts for the high efficiency of product
formation. These cosubstrates do not bind directly to the
active site of the cyclooxygenase, however, because indomethacin
(which does bind to the cyclooxygenase site) did not alter the
1:1 stoichometry of the product yield (28).

A representative example of the post-enzymatic cooxygenation
of an organic substrate via a radical chain-propagation sequence
has recently been described by Reed et al. (29). Marnett (30)
found that the synthase apparently directly oxidizes the drug
phenylbutazone by withdrawal of an electron (described more
completely in the next section), after which the electron-
deficient phenylbutazone radical combines with molecular oxygen
in solution to form a transient peroxy radical of phenylbutazone.
Subsequently, this intermediate transfers one of these oxygen
atoms to a second organic cosubstrate, thereby resulting in the
epoxidation of 7,8-dihydroxy-7,8-dihydrobenzo(a)pyrene and the
formation of 4-hydroxyphenylbutazone (29). Although it is not
established whether aromatic amines may mediate a similar
process, Boyd et al. have reported that oxygen insertion
apparently also occurs during the synthase-mediated metabolism
of the carcinogen 2-aminofluorene (31). 2-Nitrofluorene was
detected as an organic-extractable product, indicating that a
sequential series of oxygen insertion reactions had occurred.
The synthetic N-hydroxy and nitroso analogs also produced
2-nitrofluorene when substituted for the amine in identical
incubations, although these discrete intermediates were not
detected in incubations with the amine itself as the cosubstrate.
The molecular mechanism of these reactions was not elucidated,
and therefore the source of the oxygen atoms that were added to
the nitrogen atom was not identified, nor was it established
whether the amine was bound to the enzyme during the insertion
reaction(s). However, substantial inhibition of product formation by phenylbutazone suggests that both cosubstrates may be
participating in a post-enzymatic radical-propagating process
similar to that studied by Reed et al. (29). Therefore, it was

suggested that N-hydroxylation of an enzyme-bound amine, such as
is known to occur with cytochrome P-450, may not occur during
the oxygenation of 2-aminofluorene by prostaglandin synthase.
Boyd et al. also quantitated the amount of the cosubstrate that
covalently reacted with protein, or that formed other water
soluble metabolites, and in addition characterized other organic
extractable metabolites which may have been formed by
dimerization of electrophilic radicals (31). The latter are
also characteristic of metabolism by peroxidases whereby the
reducing equivalents are provided by electron withdrawal from
the substrate, in which an oxygen insertion step is not
involved. This process is described in the next section.

2.2. Cooxidation by electron withdrawal

As described in a previous section, the mechanism of the
generalized catalytic cycle of hydroperoxidase enzymes is
believed to include a highly oxidized perferrylheme intermediate
(17,20). Since an "oxidative" reaction may be defined as either
incorporation of an oxygen atom into the substrate or as a
removal of one or more electrons from the substrate, it now
appears that some of the substrates known to be metabolized by
prostaglandin synthase may be oxidized directly via the latter
mechanism. Many such substrates have been studied by organic
electrochemists and are known to yield reactive oxidation
products following anodic removal of electrons (32-34).
Experimental evidence is slowly accumulating which indicates
that corresponding oxidation products may be created
enzymatically and subsequently released into solution by a
mechanism that does not include a prior oxygen insertion step.
In this way, the oxidized enzyme intermediate may gain the
reducing equivalents required for the return to its initial
state. In this section, the evidence that this process occurs
is presented and compared for several of the most important and
well-characterized cosubstrates that have been studied to date.

2.2.1. Acetaminophen. Because of its widespread use as a
mild analgesic, the metabolism of acetaminophen (N-acetyl-p-
aminophenol; Tylenol) has received close scrutiny. In contrast
to the bladder carcinogens described in subsequent sections,

acetaminophen is an hepatocarcinogen at high doses in mice (35). Nonetheless, the metabolism of this drug is relevant because it is currently the most clearly understood example of an aromatic amine derivative that has been shown to be metabolized by a hemoprotein enzyme in a process that does not include N-hydroxylation. Several representative animal models and many of the biochemical events that occur following acute overdose have been developed and elucidated by J. R. Mitchell, J. A. Hinson, and other co-workers of J. R. Gillette, and have been reviewed recently (36). These studies have shown that hepatic cytotoxicity is mediated by an electrophilic metabolite of acetaminophen that is formed by the cytochrome P-450 family of mixed-function oxidases. Although less is known about the oxidative metabolism of acetaminophen by enzyme systems outside of the liver, the available evidence indicates a close similarity between the hepatic reactive intermediate(s) and those that are formed by enzymes in other tissues. Following acetaminophen overdose, acute renal failure often accompanies liver damage (37). An animal model of this condition demonstrated that renal glutathione levels become depleted and covalent attachment of radiolabelled drug occurs following a single large dose (38). Experiments with whole kidney microsomes showed that the covalent attachment required NADPH and oxygen, indicating at least partial involvement of cytochrome P-450. A subsequent study found that isolated, perfused rat kidneys produce small amounts of acetaminophen sulfate and glucuronide (the major metabolites found in vivo) and detected a mercapturic acid conjugate that apparently arose via enzymatic oxidation of the drug followed by rapid conversion of an initial glutathione conjugate (39). Taken together, these results strongly suggest the formation and subsequent reaction of the quinoneimide/semiquinone couple described below as oxidative metabolites in the kidney.

Unfortunately, the above studies failed to consider the differential distribution of enzyme systems within various regions of the kidney. It is now known that the renal cortex and outer medulla also contain several isozymes of cytochrome P-450, as does the liver (40). Significantly, no P-450 activity has been detected in the renal inner medulla (40), which is the

area of greatest nephropathy following chronic analgesic abuse
(41). Because of the known occurrence of prostaglandin synthase
in this tissue, several investigators have studied the metabolism
of acetaminophen and several close structural analogs, including
p-aminophenol (42) and p-phenetidine (43) by preparations of
this enzyme and other "model" hydroperoxidase enzymes (44).
Moldeus and Rahimtula found that both arachidonic acid and
linolenic acid hydroperoxide (an analog of PGG_2) mediated the
formation of the glutathione conjugate in incubations containing
acetaminophen, glutathione, and microsomes from sheep seminal
vesicles (45). Mohandas et al. reported that covalent attachment
of acetaminophen to protein by NADPH-dependent (P-450-catalyzed)
metabolism was highest in microsomes from the cortex, less in
those from the outer medulla, and minimal in those from the inner
medulla of rabbit kidney. The reverse results were achieved when
arachidonic acid was substituted for NADPH, and although the
covalent reaction in both situations was inhibited by glutathione
and ascorbic acid, indomethacin or aspirin inhibited only the
arachidonic acid-dependent reaction (25). Similar results were
reported by Boyd and Eling (26). In a follow-up study, Moldeus
et al. (46) surmised that the ultimate reactive metabolite(s) of
acetaminophen formed by prostaglandin synthase-containing
medullary microsomes were the same as those formed by liver
microsomes and NADPH. The reasons for this conclusion included
the formation of the same glutathione conjugate, the binding of
the intermediate(s) to protein, and the detection of a radical
species in both preparations. In addition, they observed a rapid
oxidation of glutathione to the disulfide, indicating that some
of the non-enzymatic oxidation-reduction reactions described
below may also be mediated by acetaminophen in the arachidonic
acid/prostaglandin synthase system as well (46).

Early studies of toxic aromatic amines and amides led many
investigators to believe that hepatic P-450-mediated
N-hydroxylation was involved in the activation of acetaminophen.
This reaction had previously been shown for the structural
analogs acetanilide and phenacetin (47). It was thought that
N-hydroxyacetaminophen would rapidly undergo a non-enzymatic
dehydration to yield N-acetyl-p-quinoneimine, an electrophile

(48) that has also been produced by the direct two-electron oxidation of acetaminophen (49). Attempts to detect the formation of N-hydroxyacetaminophen showed that the compound could form via enzymatic deethylation of N-hydroxyphenacetin (50), but not from acetaminophen itself, yet more covalent attachment to microsomal protein was observed with acetaminophen than with N-hydroxyphenacetin (51). These results, and others obtained in vivo (52), effectively eliminate the possibility that N-hydroxylation is involved in acetaminophen toxicity. Nonetheless, strong evidence remains which implicates the involvement of either the quinoneimide or the intermediate semiquinone radical (the one-electron oxidation product of acetaminophen). These species apparently exist in a complex equilibrium with each other such that the presence of one guarantees the presence of the other (53), as also described below for benzidine. N-acetyl-p-quinoneimine produced from acetaminophen by quantitative electrochemical conversion was found to react with sulfhydryl nucleophiles to yield equivalent amounts of the thioether conjugates that are also produced metabolically in vitro and in vivo (49). The quinoneimide is also reduced to acetaminophen by ascorbic acid (49); the latter also inhibits binding of the reactive metabolite(s) to protein in vitro (36) but apparently not in vivo (54). In addition, the quinoneimide undergoes hydrolysis to yield p-benzoquinone and acetamide (49), in agreement with the in vitro metabolism of acetaminophen (36).

Several mechanisms whereby one or more enzymes directly create either the acetaminophen semiquinone radical and/or the quinonemide have been proposed. The direct production of the quinoneimide by an intermediate cytochrome P-450 complex has been suggested by Hinson et al. (36) and Nelson et al. (44), as depicted in Figure 2. In this mechanism, two electrons are transferred to the enzyme-O_2-substrate complex in the normal way by the NADPH-cytochrome $\underline{c}$ reductase system. As one oxygen atom is reduced to water, a transitory bond forms between the unshared pair of electrons of the amide nitrogen and the remaining, reactive Fe-bonded oxygen atom. A similar intermediate has been proposed to decompose by cleavage of the Fe-O bond and result in insertion of the oxygen into the substrate

(21). With acetaminophen as the substrate, however, the N-O
bond breaks such that the electrons remain with the oxygen
(subsequently also released as water), and the electron-
deficient acetaminophen is released as the quinonemide.
According to this scheme, molecular oxygen is reduced to two
molecules of water, with two reducing equivalents derived from
the reductase system and two from the acetaminophen molecule.
If this mechanism is also a representation of the events that
might occur during the oxidation of acetaminophen by peroxidases,
it illustrates why an N-hydroxy metabolite need not be an
obligate intermediate.

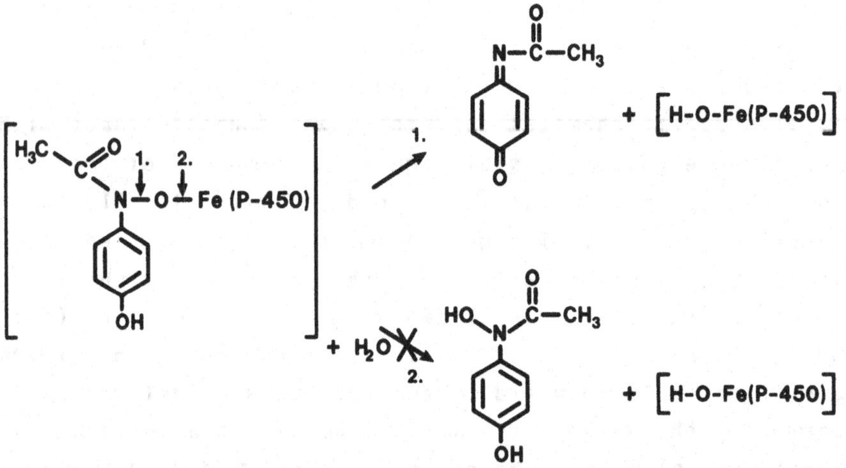

FIGURE 2. Postulated molecular mechanism for the two-electron
oxidation of acetaminophen (36).

Once formed, it has been widely believed that the reactive
metabolite(s) of acetaminophen (either the semiquinone or the
quinoneimide) cause cellular damage by covalently reacting with
and disrupting the function of critical proteins (55). According
to this view, the reactive intermediate adds faster to
glutathione than to protein, and hence formation of the conjugate

protects the cell from damage. More recently, however, investigators have realized that the enzymatic creation of the semiquinone and/or quinoneimide by either mechanism described above may initiate a cyclical sequence of non-enzymatic reactions whose cumulative influence may be far more damaging than the mere arylation of protein sites. Consequences of these reactions may include decreases in the normal levels of molecular oxygen, ATP, NAD(P)H or glutathione which may exert deleterious effects from which the cell may not recover (52-58). Reaction cycles involving this large-scale transfer of electrons mediated by chemicals such as acetaminophen could occur before the radio-labelled substrate binds to protein (the most commonly used "index" of acetaminophen metabolism). The relative rates of most of these reactions are not known, thus the importance of each remains to be established. However, it is becoming increasingly clear that much more than covalent binding to macromolecules may be involved when acetaminophen is oxidized to a reactive metabolite, and this may be true for many other toxic aromatic amines as well.

2.2.2. Benzidine. In the late 1970's, the concept of target tissue metabolism of site-specific carcinogens had received very little attention relative to the extensive studies of metabolic pathways by enzymes in the liver. Consequently, hypotheses to account for the appearance of tumors at sites distant from the liver included one or more obligate, initial metabolic transform-ations mediated by liver enzymes. It was rationalized that a re-latively stable "proximate carcinogenic metabolite" of the parent procarcinogen was then transported to the susceptible organ, at which some critical additional event(s) occurred (59,60). Strong evidence supports the likelihood that at least a portion of the metabolism of many carcinogens occurs by this pathway.

More recently, the characterization of the cooxidative metabolism of benzidine mediated entirely by enzymatic activity of the target organ and adjacent tissues grew out of initial studies of prostaglandin synthesis and of the differential distribution of drug metabolism in various regions of the rabbit kidney (61,62). As mentioned in previous sections, these investigations revealed that appreciable levels of cytochrome

P-450 exist only in the renal cortex, while prostaglandin synthase activity is low in the cortex, higher in the outer medulla and highest in inner medulla. Incubations of radiolabelled benzidine with microsomes prepared from outer or inner medulla and various other components showed that radioactivity became covalently bound to DNA, tRNA, protein, and also to metabolites that remained in the aqueous fraction after complete removal of unreacted benzidine by organic extraction. This conversion was dependent upon addition of one of several fatty acid precursors of prostaglandins, or hydrogen peroxide, or other organic peroxides, but not upon other fatty acid substrates. Indomethacin and aspirin, which are known inhibitors of the cyclooxygenase activity that generates prostaglandin hydroperoxide intermediates, were found to inhibit the activity initiated by the fatty acid precursors of prostaglandins but not that of the peroxide substrates. Conversely, conversion of benzidine to bound products was not observed following addition of NADPH to medullary microsomes, nor in incubations containing hepatic or renal cortical microsomes with either NADPH, arachidonic acid, or peroxides (63,64). A follow-up investigation focused specifically on the activity of the corresponding canine tissues and included microsomes from bladder epithelial tissue for comparison (11). The conversion of benzidine to aqueous-soluble metabolites was four-fold greater with bladder microsomes compared to inner medullary microsomes when initiated by arachidonic acid, as illustrated in Figure 3. This activity was not observed with the addition of NADPH. In view of these observations, it has been concluded that the hydroperoxidase activity of prostaglandin synthase is responsible for the conversion of benzidine to reactive intermediates. Additional work showed that such peroxide-initiated activity was not unique for this enzyme, but could also be seen with several other peroxidase enzymes (65-67). As described in the preceding section, acetaminophen is also known to be cooxidized by various peroxidases, resulting in covalent attachment to protein and the formation of a glutathione conjugate (36). Although the molecular picture of these events is far from complete, the mechanism of oxidation of benzidine as it is now envisioned appears to be much more closely related to that described for acetaminophen than to the classical mechanism of hepatic N-oxidation that is generally accepted for many other aromatic amines. In particular, no evidence for N-hydroxylation during peroxidatic cooxidation of benzidine has been reported. Thus, if the analogy with acetaminophen is valid, an oxidized intermediate in the enzymatic cycle directly withdraws one or two electrons from the benzidine molecule, thereby creating either the corresponding semiquinoneimine or the diimine as electrophilic metabolites.

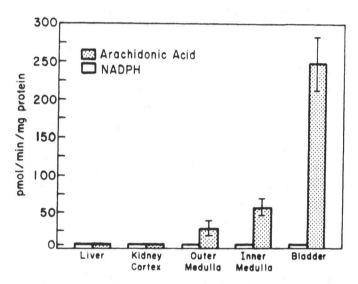

FIGURE 3. Aqueous-soluble products of benzidine metabolism formed by microsomes from various canine tissues (11).

Numerous additional experimental observations support the above mechanism. Synthetic N-hydroxybenzidine is reported to be very unstable (68), and probably decomposes via dehydration to yield the diimine in a manner analogous to that invoked for synthetic N-hydroxyacetaminophen (69), but at a faster rate. Thus, if an initial Fe-O-N complex exists at the active site, the decomposition/rearrangement could occur while the molecule remains bound to the enzyme. The net result would be the withdrawal of two reducing equivalents from the benzidine molecule and release of a water molecule and the diimine by the resting-state enzyme. Benzidine is known to be electrochemically oxidized in a two-electron process at a potential distinctly lower than that seen for acetaminophen (70,71), indicating the relative ease of the electron withdrawal reaction. Rice and Kissinger (65) found that chemically-synthesized benzidine-diimine reacted with various sulfhydryl-containing nucleophiles to produce putative thioether conjugates analogous to those known for acetaminophen (36,49). In addition, an identical chromatographic sequence of products was observed following the reaction of the diimine with glutathione at physiological pH, and also when a corresponding solution of benzidine and glutathione is cooxidized by horseradish peroxidase and hydrogen peroxide. This suggests the involvement of the same reactive intermediate in each situation. It is also known that the synthetic diimine is reduced to benzidine by ascorbic acid (65,72), which also prevents the attachment of radioactivity to protein following cooxidation of radiolabelled benzidine by prostaglandin synthase (72).

An additional element in the analogy between the oxidation of benzidine and acetaminophen by peroxidases is the experimental identification of the semiquinoneimine cation radical. Although the exact identity of the acetaminophen-derived radical has not been determined (44,52), the corresponding situation with benzidine has been much more clearly defined. Josephy et al. have examined the peroxidation of the benzidine analog 3,5,3',5'-tetramethylbenzidine (TMBD) by horseradish peroxidase (73) and prostaglandin synthase (74) and observed a radical signal by ESR spectroscopy. Comparison of the hyperfine splitting of this spectrum with a computer simulation, made using literature values as likely nuclear splitting constants, established the identity of the semiquinonediimine cation radical. This also indicates that the radical exists free in solution rather than bound to the enzyme. At acidic pH, this entity is stable for several hours, while at neutral pH only an unresolved line spectrum was seen. These workers also postulated that, in incubations with submolar amounts of peroxide relative to TMBD, the semiquinoneimine rapidly established an "equilibrium" with a blue charge-transfer complex comprised of one molecule of TMBD and one of the TMBD

diimine. With equimolar or greater amounts of peroxide, only
the diimine was detected. Previously, Claiborne and Fridovitch
(75) had suggested that a similar relationship existed during
the peroxidation of o-dianisidine (3,3'-dimethoxybenzidine),
although they failed to detect a radical in this situation.
This species was successfully detected as part of the TMBD study
(73). A mathematical analysis of this situation employing
absorbance measurements of the principle components revealed
that the steady-state concentration of the semiquinoneimine is
negligible compared to that of the diamine/diimine complex (73).

The preceding studies of benzidine analogues were extended to
benzidine itself by Wise et al. (76), who established that an
identical ESR spectrum was obtained during peroxidation of
benzidine with either horseradish peroxidase or prostaglandin
synthase, or by computer simulation. They also showed that the
characteristic time course of appearance of the radical signal
could be interrupted by the addition of KCN (a potent,
irreversible inhibitor) or acetaminophen and abolished
completely by ascorbate. The completely resolved spectrum of
the benzidine semiquinoneimine cation radical product by
prostaglandin synthase is shown in Figure 4.

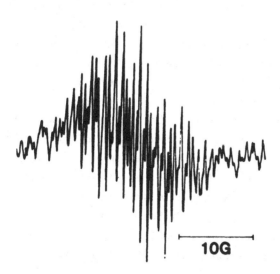

10G

FIGURE 4. Electron spin resonance spectrum of the
semiquinonediimine of benzidine produced during the oxidation of
benzidine by prostaglandin synthase (76).

To summarize the observations that have been published to date about the solution chemistry of the oxidized products of benzidine and its derivatives, peroxidatic reactions in which the oxidant is limiting have yielded at least four solution components: the diamine, the diimine, the charge-transfer complex, and the semiquinoneimine radical. As with the acetaminophen situation, the question of whether the semiquinoneimine or the diimine represent the initially-released enzymatic oxidation product has not been resolved.

A unified scheme that encompasses all of the observations described above has recently been published (72) and is shown in Figure 5. These processes depicted are not unique for benzidine and may be generalized to include all compounds that are peroxidized by prostaglandin synthase. An important feature of this diagram is the specification of four distinct points at which the process resulting in the binding of the oxidized benzidine intermediate(s) to cellular macromolecules can be prevented. It has been suggested that administration of pharmacological agents that interver. at one or more of these sites may attenuate the initiation of the carcinogenic process (72).

At Site 1 of Figure 5, agents that prevent the formation of hydroperoxide substrates of prostaglandin synthase or that prevent their interaction with the enzyme thereby block the entire hydroperoxidase reaction. This inhibition is designated as occurring at site 1a if it specifically involves the generation of the hydroperoxide from a fatty acid precursor by prostaglandin synthase itself. Prevention of benzidine oxidation by aspirin and indomethacin (12,63) thus occurs at site 1a. Experimental evidence for the in vivo inhibition of prostaglandin production by the transitional epithelium of the urinary bladder following 0.5% aspirin feeding has recently been published (12). Site 1b designates the various means whereby formation of any other suitable hydroperoxide substrates, such as a non-eicosinoid lipid hydroperoxide or H_2O_2, formed by enzymes or processes other than the cyclooxygenase activity of prostaglandin synthase, are prevented.

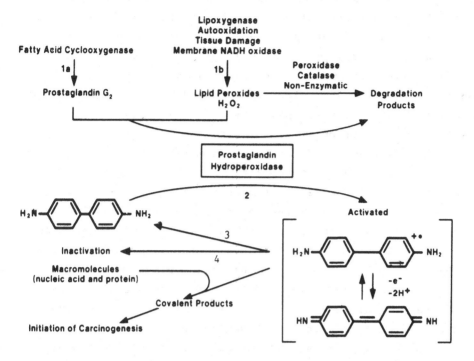

FIGURE 5. Formation and fate of the oxidized intermediate(s) of benzidine catalyzed by prostaglandin synthase (72).

Alternatively, the oxidation reaction could be blocked or greatly diminished despite the normal interaction of hydroperoxide with the enzyme, and agents that cause this to occur operate at site 2. In so doing, these agents may themselves become oxidized by either oxygen insertion or electron withdrawal, or they may simply block the active enzyme site and prevent the necessary benzidine-enzyme interaction. Consequently, many of the agents found to be effective at Site 2 are non-toxic cosubstrates of the synthase and probably act by competitive inhibition. As examples, propylthiouracil, methimazole and phenidone were found to diminish binding of benzidine to protein. The first two drugs are used clinically as thyroid peroxidase inhibitors. Interestingly, methimazole has been found to have no effect on the binding of 2-aminofluorene to protein when catalyzed by hepatic mixed-function oxidases (77).

This information is important because whole-animal studies with these inhibitors could help in delineating the relative contribution of various competing metabolic pathways to the overall tumorigenicity of test carcinogens. Moreover, their use as part of a therapeutic regimen for individuals suspected of chronic exposure to environmenal carcinogens could aid substantially in diminishing the initiation phase of the carcinogenic process.

Agents that function at sites 3 and 4 of Figure 5 intercept and chemically react with the reactive metabolite(s) of benzidine after the procarcinogen has been oxidized by the synthase, but before reaction with a cellular macromolecule occurs. Site 3 specifies a chemical oxidation/reduction reaction in which a cytosolic reducing agent provides electrons to reduce the benzidine metabolite back to unaltered benzidine, thereby accomplishing the reverse of the reaction that takes place between the enzyme and the procarcinogen. In so doing, the reducing agent itself becomes oxidized. This reaction is analogous to many of the non-enzymatic solution reaction that were described in the previous section involving acetaminophen, but which could also occur with benzidine. In vitro, the most studied example of a site 3-acting species is ascorbic acid (vitamin C), which has also been reported to be a substrate for horseradish peroxidase and thus could also act at Site 2 (78). This agent inhibits benzidine binding to macromolecules (72), quenches the ESR signal of the benzidine cation radical (76) and quantitatively converts synthetic benzidinediimine to benzidine (72). An important consequence of the occurrence of this reduction process is that the unaltered benzidine that results cannot be distinguished from benzidine that was not metabolized at all. Thus, a considerable amount of "redox cycling" (also called "non-productive metabolism") of the benzidine molecules could occur before an observable alteration in the molecule took place, and this could result in an underestimation of the rate or total amount of metabolism that has occurred (65,72).

Reactions that occur at site 4 produce cytosolic benzidine-derived metabolites that do not react with macromolecules, and which thus reflect a detoxification of the oxidized reactive intermediate(s). As reaction products, the structures of these metabolites are of interest because the mechanism of their formation may mimic that which occurs with critical macro-molecules. At least two types of such post-oxidation reactions have been reported with benzidine as the oxidized substrate. As described above, thiol-containing nucleophiles react with enzymatically- or chemically-oxidized benzidine to yield thioether conjugates. This suggests that the endogenous nucleophile glutathione may function in vivo to detoxify the reactive intermediate(s) of benzidine as it is known to do with many other compounds. In addition to operating in this way at site 4, such nucleophiles also act at site 3 and thereby become oxidized to disulfide products in the process (65). Thus, glutathione may also contribute to the oxidation/reduction cycling process described above.

A second class of benzidine-derived reaction products has been studied recently by Josephy and coworkers (74,79,80) and involves the coupling of one of the nitrogen atoms of an oxidized benzidine intermediate to a suitable phenolic compound at the position para to the hydroxy group, resulting in the formation of a substituted indophenol product. Phenolic compounds having an easily-displaced substituent in this position also react. Thus, butylated hydroxyanisole (BHA) couples with oxidized benzidine while butylated hydroxytoluene (BHT) fails to yield a product. This reaction pathway is of interest because of the known inhibitory effect of certain commonly-used dietary phenolic antioxidants, including BHA, on the metabolism of certain chemical carcinogens and upon the sub-sequent appearance of tumors (81,82). As does glutathione, it is conceivable that such antioxidants could directly exchange electrons with the benzidine intermediate(s) (site 3 activity) as well as covalently reacting. Furthermore, if these phenolic agents are themselves oxidized directly by the enzyme, they could represent very effective inhibitors of benzidine-induced carcino-genesis by simultaneously inhibiting the process at sites 2, 3, and 4 of Figure 5.

The preceding description of prostaglandin synthase-mediated metabolism dealt exclusively with benzidine as an otherwise unaltered procarcinogen. However, since relatively high levels of the synthase exist in the tissues of the urinary tract (the site of benzidine-induced tumors) (11,12), an additional impor-tant factor in the overall carcinogenic process involves the effect of prior metabolic transformations that occur before the resulting benzidine metabolites reach these tissues. Recent evidence suggests that the extent to which benzidine is metab-olized in the liver exerts a profound effect upon the eventual site of tumor formation. For example, rodents appear to be primarily susceptible to benzidine-induced hepatic carcinomas, while dog and man are susceptible to aromatic amine-induced urinary bladder cancer. This fundamental difference appears to correlate with the capacity of the species for hepatic N-acetylation of these compounds. In certain species, notably man and the rabbit, but not the rat, the rate of N-acetylation of amine-containing drugs and of numerous aromatic amine carcinogens is subject to a genetic polymorphism that has been attributed to differences in the amount of N-acetyltransferase activity in the liver and in other organs (83 and references therein). Because of this polymorphism, members of the relevant species are distributed bimodally into "rapid" and "slow" acetylator populations, and these differences have been closely correlated with the degree of susceptibility to drug-induced toxicity. This has led to the suggestion, as yet unproven, that human slow acetylators may be at greater risk of aromatic amine-induced urinary bladder cancer (84). By contrast, dogs express very little N-acetyltransferase activity, and hence do not appreciably acetylate the amine functions of foreign compounds (85). This appears to correlate with the finding that dogs develop tumors of both liver and bladder upon administration of acetylated aromatic amines, but only bladder tumors when the parent amine is given. This suggests that the presence of an acetyl group is required for initiation of carcin-ogenesis by the process(es) that operate in the liver, but may have less effect upon those that are important in the urinary tract. Additional biochemical evidence for this view has been provided by detailed studies of the pathway leading to formation of the major benzidine-DNA adduct in rat or mouse liver, which most probably involves N-acetylation followed by cytochrome

P-450-catalyzed N'-hydroxylation (i.e., oxygen insertion at the non-acetylated amine group) (68). O-Esterification followed by dissociation apparently yields the electrophilic entity that binds to the DNA to yield N-(deoxyguanosine-8-yl)-N'-acetylbenzidine. Neither N-hydroxy-N-acetylbenzidine nor N,N'-diacetylbenzidine were implicated in this process (68). Presumably, this adduct would not be seen in dog liver.

The acetylated derivatives of benzidine are reported to be metabolized much more slowly by preparations of prostaglandin synthase than is benzidine itself (67,86). It is noteworthy that the electrochemical oxidation potential of benzidine (which reflects the energy required to withdraw electrons) increases by ca. 200 mV following each sequential acetylation of the amine groups (71). It appears that hydroperoxidase enzymes have only limited ability to withdraw electrons from cosubstrates. Thus, this process occurs very quickly for those amines which have a low oxidation potential (e.g., benzidine), but at a much slower rate for the corresponding monoacetylated (or diacetylated) analogue. The acetylated forms would be correspondingly more stable while positioned adjacent to the highly-oxidized enzyme intermediate and thus more amenable to the radical-recombination reaction that is believed to occur during oxygen insertion (21). Hence, the latter process (hydroxylation, followed by O-conjugation) becomes the more favored route for eventual formation of an electrophilic metabolite from most acetylated amines. On the other hand, both mechanisms may be equally facile for some substrates, an example of which is presented in the following section. If this view is correct, N-acetylation in the liver and elsewhere may represent a protective mechanism with regards to subsequent activation by peroxidase-rich tissues, while simultaneously rendering the molecule more susceptible to activation via hydroxylation.

2.2.3. 4-Aminobiphenyl. Although the most complete molecular picture of the prostaglandin synthase-catalyzed mechanisms of aromatic amine cooxidation has been obtained from studies of acetaminophen and benzidine, numerous other aromatic amine substrates have also been examined (86,87). Of particular interest among these are 2-naphthylamine and 4-aminobiphenyl,

which are believed to be stronger carcinogens for the urinary
bladder than is benzidine (4,5,60). Previously, molecular
oxidation mechanisms involving hepatic N-hydroxylation and
subsequent esterification were advanced as feasible explanations
of the activation process (88,89) for these molecules. With
4-aminobiphenyl, this reaction may be experimentally achieved in
vitro by incubation of N-hydroxy-4-acetylaminobiphenyl with
N,O-acyltransferase, which intramolecularly shifts the acetyl
group from the nitrogen atom to the oxygen atom. Heterolytic
cleavage of the N-O acetoxy bond yields the nitrenium ion of
4-aminobiphenyl, a strongly electrophilic entity that is
deficient by two electrons relative to the parent amine (90).
This reactive intermediate is known to react preferentially at
the C-8 position of deoxyguanosine (91). However, a recent
study by Morton et al. (86) compared the specificity of this
reaction with the reactivity of the prostaglandin synthase-
catalyzed reactive intermediate of 4-aminobiphenyl towards
various nucleosides. They found preferential reactivity towards
polycytidylic acid rather than towards polyguanylic acid. This
significant difference in reactivity indicates that different
reactive intermediates are involved, and consequently the
authors concluded that the prostaglandin synthase-catalyzed
product was not the nitrenium ion (86). If, however, the amine
were being oxidized by removal of a single electron, an
electrophilic radical might be the product which would,
therefore, display reactivity different from that of the
nitrenium ion and thus bind in a different manner to nucleic
acids. Although far from conclusive, this work is further
suggestive evidence that a hydroperoxidase may oxidize
monoarylamines by an electron withdrawal mechanism that does not
involve prior oxygen insertion. It should remain of continuing
interest to develop selective inhibitors of each of these
complementary metabolic pathways so that the relative
contributions of each may be elucidated in vivo.

2.2.4. Aminothiazole-substituted 5-nitrofurans. The
suggestion that prostaglandin synthase was involved in the
initiation of bladder cancer subsequently led to the examination
of the effect of this enzyme on certain other well-characterized

model bladder carcinogens. The carcinogenicity of the more common, industrially-produced aromatic amines described in the previous sections was initially suspected following occupational exposure to humans. By contrast, far more potent and specific bladder carcinogens have been synthesized and are routinely in use in studies of the two-stage etiology and progression of this disease, primarily in rodent models. Prominent among these are certain heterocyclic compounds which have in common a ring-substituted 5-nitrofuran moiety. Much of the work in this area, and a comprehensive review of its status, have been provided by G. T. Bryan and coworkers (92). Among the most extensively studied members of this class are those that bear a substituted thiazole ring, and of particular interest are the analogs with a free or acylated amine group in the 2-position of this ring. The structures of three of these are shown in Figure 6. The free amine analog is 2-amino-4-(5-nitro-2-furyl)thiazole (ANFT), while formylation of the amine yields N-[4-(5-nitro-2-furyl)-2-thiazolyl]formamide (FANFT) and acetylation yields N-[4-(5-nitro-2-furyl)-2-thiazolyl]acetamide (NFTA). These small variations in structure result in greatly differing species and tissue susceptibility towards induction of tumors. FANFT very specifically induces bladder cancer in rats, mice, hamsters, and dogs, with similar results observed in males and females. The resulting bladder neoplasia has histologic characteristics similar to those reported for humans. Rats fed the compound at 0.2% of dietary weight for 10 or more weeks develop a 70-100% incidence of irreversible lesions that eventually result in carcinoma (93). Feeding of ANFT to mice or rats produces primarily forestomach tumors, with lesser numbers of tumors appearing in several other tissues, including the bladder (94,95). NFTA causes tumors of the renal pelvis in the rat (96) and bladder cancer in the hamster (97). Although it is clear that such tumorigenic activity at sites remote from the point of administration suggests the requirement for end-organ metabolic activation, very little is known about the enzymatic activities actually responsible. On one hand, whole-animal structure-activity studies have shown that the characteristic bactericidal, mutagenic, and carcinogenic properties of a number of

5-nitrofurylthiazole analogs are lost upon deletion of the nitro group. This has led to extensive investigations of activation processes involving nitroreduction (98-100), which may be required for some, but perhaps not all, of the aforementioned biological effects. On the other hand, coadministration of FANFT and allopurinol, a nitroreductase inhibitor, increased rather than decreased the incidence of bladder tumors in rats (101). In addition, it is difficult to rationalize how acylation of a distal amine function could exert enough effect upon the reduction process to so greatly affect the distribution of responses among these compounds. Thus, it remains possible that absence of the nitro group has a greater impact on the absorption, distribution, and alternate detoxification pathways than it has on the activation process, and that other activating enzymes besides nitroreductases are involved.

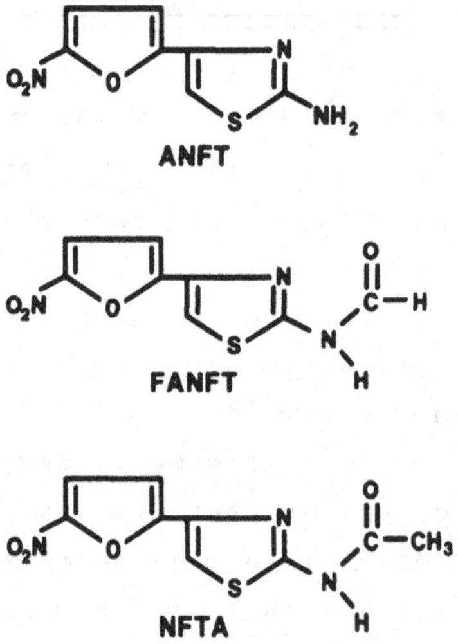

FIGURE 6. Structures of three aminothiazole-substituted 5-nitrofuran carcinogens.

In order to investigate the possibility that certain of the 5-nitrofurans are cooxidatively activated by prostaglandin synthase, the relative amounts of radiolabelled ANFT and FANFT that became covalently bound to cellular macromolecules was determined (102). Microsomes containing the synthase were prepared from ram seminal vesicle, dog bladder transitional epithelium and rabbit renal inner medulla, and each gave similar results. As with the cooxidative mechanisms described previously with the other aromatic amines, substrate and inhibitor experiments implicated the hydroperoxidase component of the synthase. Moreover, oxygen was required for metabolism initiated by arachidonic acid but not by peroxide-containing substrates. As shown in Table 1, the rate of the covalent attachment of ANFT to protein is at least 4-fold greater than that of FANFT. Significantly, previous studies have indicated that FANFT is substantially deformylated in vivo (100) and by various tissue

Table 1*. Solubilized ram seminal vesicle prostaglandin synthase-catalyzed binding of FANFT and ANFT to protein and DNA.

Metabolism of either 0.05 mM [^{14}C]FANFT or [^{14}C]ANFT was initiated by addition of 0.06 mM arachidonic acid to 0.44 mg/ml solubilized microsomal protein in 100 mM phosphate buffer, pH 7.8.

	Protein bound	DNA bound
	nmol/mg protein/min	
FANFT	0.2 + 0.01	0.002 [†]
ANFT	0.9 + 0.03 [‡]	0.05 + 0.01 [‡],[§]

* Taken from ref. (102)
† Limit of detection
‡ $P < .01$ compared to corresponding value for FANFT
§ Mean + S.E. with n = 6

preparations in vitro (103), and that the mutagenicity of urine from FANFT-fed rats correlates with the concentration of ANFT rather than FANFT. Thus, the cooxidation of ANFT by prostaglandin synthase may be of central importance in FANFT-induced bladder cancer. The lower rate of FANFT cooxidation relative to ANFT was also reflected in the observed extent of binding of each to DNA (Table 1), such that no binding of FANFT was detected under the conditions selected for the assay. NFTA also displays a lower degree of reactivity similar to that of FANFT (104).

Although the radiolabelled binding studies described above clearly reveal that metabolic conversion of these aminothiazoles has taken place, the molecular mechanism of this transformation and the structures of the metabolites that result have not been

elucidated, particularly because of the inherent instability of
the two heterocyclic rings and the presence of two reactive
functional groups. However, preliminary evidence suggests that
the oxidative metabolites of ANFT are different from those
resulting from nitroreduction (99,102). Consequently, by
analogy with the known products of prostaglandin synthase
metabolism of other aromatic amines, it is necessary to consider
the possibility that the oxidation reaction could occur by either
oxygen insertion or electron withdrawal. Evidence for the former
possibility has been obtained by the tentative characterization
of an organic-extractable metabolite of ANFT having an additional
oxygen atom in the 4-position of the furan ring that was isolated
from an incubation containing solubilized microsomal
prostaglandin synthase (102). As indicated earlier, this new
substituent may have originated from the peroxide or directly
from molecular oxygen. On the other hand, recent electrochemical
experiments have indicated that ANFT is oxidized at ca. +800 mV,
(J.R. Rice, unpublished observations) which is approximately the
same energy required for oxidation of the aromatic amines aniline
and 4-aminobiphenyl (ca. +700 mV) (34,105). By contrast, no
oxidative electrochemical process is observed with either FANFT
or NFTA at applied potential below +1200 mV (the solvent limit
in aqueous solution), in agreement with the known effect of
acylation of primary aromatic amines upon their subsequent
electrochemistry (34). These observations suggest that an
electron withdrawal mechanism analogous to that described for
acetaminophen and benzidine may also be feasible for amino-
thiazole derivatives, such that the rate of electron withdrawal
by the synthase is proportional to the oxidation potential of
the cosubstrate. This would account for the difference in
metabolism of FANFT and NFTA relative to ANFT shown in Table 1
and for the observation that non-amino-substituted analogues of
ANFT do not appear to be substrates for cooxidation by prostaglandin synthase (M.B. Mattammal, unpublished observations). In
conclusion, the extreme potency and inherent chemical reactivity
of these carcinogens indicate that several activation mechanisms
leading to different but effective reactive intermediates may be
possible, and that these may include both oxygen insertion or

electron withdrawal by prostaglandin synthase. As with all of the other molecules oxidized by this enzyme, intervention at the sites specified in Figure 5 could modify or inhibit this process.

3. IN VIVO STUDIES IMPLICATING PROSTAGLANDIN SYNTHASE IN BLADDER CANCER

One of the ultimate goals of delineating the mechanism of metabolic conversion of a chemical carcinogen in vitro is the rational development of additional experiments to establish the relative importance of this pathway in the whole animal. In contrast to the often very simple sub-tissue preparations that suffice for in vitro work, studies utilizing living, functioning mammals usually cost the investigator dearly in terms of time, money and loss of control of experimental variables. This is chiefly due to the lack of a conveniently-observed "index" of change (decrease in absorbance, covalent binding of radiolabel, production of a discrete chemical product) which usually expedite the aquisition of easily-interpreted in vitro data. Animals are frustratingly slow in revealing evidence of chronic pathological changes in their bodies, and this reality has usually led to the selection of the most potent (i.e., shortest-acting) chemicals for the study of reproducibly-induced cancer. A second consequential requirement has been the application of statistical analyses in order to mathematically demonstrate positive results and correlations, thus requiring large numbers of animals with which to establish an adequate data base. This in turn places a premium upon studies with small, easily-maintained mammals, such as rodents. Consequently, for reasons of practicality and convenience, in vivo studies of the involvement of prostaglandin synthase in the metabolic activation of aromatic amine carcinogens have employed the rat model of FANFT-induced bladder cancer (Section 2.2.4). The experimental rationale has been to administer a pharmacological agent to inhibit prostaglandin synthase activity, and to compare the FANFT-induced tissue changes that occur in this situation with those seen in the non-inhibited control case in which FANFT alone is fed.

In Section 2.2.2, several chemical approaches for attenuating the prostaglandin synthase-mediated cooxidation sequence were

described and specified in Figure 5. Of the various pharmacological agents having activity at these sites, inhibition of
arachidonic acid oxygenation at Site 1a by aspirin was selected
because of the extensive knowledge of its pharmacological effects
on the whole animal and because previous studies had shown that
aspirin feeding inhibited prostaglandin synthase (106).
Indomethacin, a second non-steroidal anti-inflammatory inhibitor
of the synthase which is commonly used with in vitro enzyme
studies, was found to be toxic when chronically fed at levels
high enough to achieve in vivo inhibition of bladder
prostaglandin production (12).

In the first of two studies reported to date, the appearance
of hyperplastic lesions of the bladder epithelium induced by
feeding 0.1% or 0.2% FANFT for 6 or 12 weeks (12) was selected
as the experimental observable. Groups of 3-8 rats were fed
either control diet, diet plus 0.5% aspirin, or diet containing
FANFT with or without 0.5% aspirin. After the 6-week period,
the rats receiving aspirin plus 0.2% FANFT had a significantly
lower number of lesions, and of a smaller size, than those
receiving only 0.2% FANFT, although the total number of affected
animals was similar. After the 12-week period, there was again
some inhibition by aspirin of the lesions induced by 0.2% FANFT,
but complete inhibition of the lesions induced by 0.1% FANFT. No
lesions were observed in rats fed aspirin without FANFT or in
control rats.

A second, longer term study examined the effect of FANFT plus
aspirin on the formation of bladder carcinoma (107). The design
of the feeding regimen employed is shown in Figure 7. One group
of rats (Group 4) received only control diet for the entire 76
weeks of the study. The animals in Group 3 received only 0.5%
aspirin, beginning two days before the initiation of FANFT
feeding to Groups 1 and 2. Of these, Group 1 received only 0.2%
FANFT for 12 weeks, followed thereafter by untreated control
diet. Group 2 received 0.5% aspirin for two days, followed by
0.2% FANFT plus 0.5% aspirin for the next 12 weeks, then the
aspirin diet for an additional week, and finally untreated
control diet for the remaining 63 weeks of the study. The
purpose of feeding the inhibitor prior to and subsequent to the

FANFT feeding period was to insure maximum inactivation of the cyclooxygenase component of the enzyme while FANFT was present in the tissues of the animals. As was found in the previous study, the results indicated a substantial decrease in the incidence and severity of the disease. The rats fed FANFT alone were found to have bladder tumors in 18 or 21 cases (87%), while those receiving both FANFT and aspirin produced tumors in only 10 of 27 animals (37%). With each of these studies, it is important to recognize the limitations attending the use of an inhibitor agent at any one of the sites depicted in Figure 5, especially those that do not act directly at Site 2. In the case of a non-toxic dose of aspirin, the inhibition occurs by attenuating the generation of the hydroperoxide PGG_2 from arachidonic acid by prostaglandin syntase itself. However, aspirin does not inhibit either the formation of any other cellular hydroperoxides by other processes, or the reduction of these by the synthase. Hence, a residual amount of cooxidation of the proximate carcinogen of FANFT or ANFT could still lead to a statistically significant appearance of hyperplasia or carcinoma among the aspirin-inhibited animals. In this regard, medullary microsomes

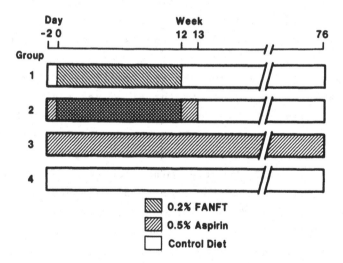

FIGURE 7. Feeding regimen for the study of the inhibition of FANFT-induced bladder carcinoma by aspirin (107).

from aspirin-treated rabbits mediated the metabolism of
acetaminophen and benzidine when initiated by a fatty acid
peroxide as well as did microsomes from control rabbits (106).
Further work is clearly necessary to devise suitable combinations of inhibitors such that the activation process may be
inhibited at multiple sites in Figure 5. Alternatively, as
pointed out in the previous section, the extreme potency that
makes this carcinogen useful in such feeding studies may also be
due to activation mechanisms involving other enzyme systems
(e.g., nitroreductases) that are not aspirin-sensitive. In this
case, whole-animal inhibitor studies such as those described
here may provide an estimation of the relative contribution of
each pathway to the overall tumorigenic process.

4. CONCLUSIONS AND FUTURE DIRECTIONS

A major goal of contemporary chemical carcinogenesis research
is the ability to predict, on the basis of molecular structure,
those compounds likely to induce tumors in humans. As this
approach becomes more feasible, the necessity of reliance upon
the expensive and imperfect long-term, massive-dose animal
feeding programs will become correspondingly lessened. A
requirement for the development of this ability is a clear
understanding of the biochemical processes that interact with
known carcinogens, and the manner in which the chemical alters
these processes and the structure and functions of the cells in
which they take place. This chapter discussed recent investigations of the interactions of prostaglandin synthase with selected
carcinogenic aromatic amines. This enzyme displays prominent
activity in the urinary bladder transitional epithelium, which
is the primary site of aromatic amine-induced tumors in humans.

The early biotransformation studies of carcinogenic aromatic
amines logically focused upon the more stable and detectable
metabolites. Ease of detection posed the requirement that a
relatively large amount of the product be isolated, which in
turn discriminated against metabolic processes that took place
in organs other than the liver. This situation led to the
identification of N-hydroxy-2-acetylaminofluorene as a proximate
carcinogen (108). Subsequent work with similar amines and

amides led at one point to the assertion "that N-hydroxy derivatives are obligatory intermediates in the carcinogenicity of aromatic amines, amides, and nitro compounds" (109). This statement now appears to be overgeneralized. Although N-hydroxylation may be a required oxidative reaction for some amine carcinogens, and although prostaglandin synthase may cause the N-oxidation of some amines to occur, it is equally clear that certain other toxic aromatic amines may exert their prostaglandin synthase-mediated effects via reaction pathways that do not include prior N-hydroxylation.

The peroxidase reaction catalyzed by prostaglandin synthase results in the reduction of a hydroperoxide to an alcohol. This reaction requires the input of two reducing equivalents, and the function of the enzyme in this process is to mediate the transfer of these equivalents, either in the form of electrons per se, or as an electron-deficient, oxygen-containing fragment that had been cleaved from the peroxide, and which has many of the chemical characteristics of the hydroxyl radical (23). The degree to which this latter entity is reduced to water (as opposed to covalently adding itself to the enzyme or another molecule) may depend upon the tendency of the amine cosubstrate to release electrons. Table 2 presents a compilation of several of the amines that were described in this chapter, along with a corresponding electroanalytically-obtained oxidation potential

Table 2. Comparative electrochemical half-wave potentials for the oxidation of selected toxic aromatic amines

Half-wave oxidation potentials were obtained by cyclic voltammetry (70) at a carbon electrode with a scan rate of 150 mV/sec. 1.0 mM analyte was present in 10:90 ethanol:0.1 M phosphate buffer, pH = 5.0.

Compound	$E_{1/2}$*	Reference
4-Aminophenol	+350	70
Benzidine	+425	105
3,3'-Dichlorobenzidine	+500	105
Acetaminophen	+550	70
4-Aminobiphenyl	+610	105
ANFT	+800	–
Phenacetin	+875	70

*mV versus Ag/AgCl reference

value. As is seen, those compounds that have been reported
(86,87) to be the most reactive in prostaglandin synthase-
containing incubations also have the lowest oxidation potentials
(benzidine, acetaminophen), while those that have been reported
to be oxygenated in these incubations (ANFT, 4-aminobiphenyl)
are representative of amines with higher oxidation potentials,
indicating that they release electrons less readily. Phenacetin
has been reported to be inert to metabolism by the synthase in
vitro (87). As a consequence of having provided reducing equivalents via electron withdrawal by the enzyme, the initial amine
oxidation products may correspond to the free radical, semiquinone, or quinoneimine metabolites that have been identified or
postulated to be formed. Worthy of additional consideration are
the secondary or post-enzymatic reactions with dissolved oxygen,
thiols, or other reactive solution constituents that these
oxidized intermediates may participate in. Because of these and
because of their extreme reactivity, it is unlikely that either
the initial enzyme-released product, or the amine-derived entity
that actually participates in the deleterious carcinogenic
reaction (which may not be the same) will ever be identified.

A logical experimental approach for confronting this problem
may be the study of the stable "end products" that are ultimately
derived from the unstable reactive intermediates, thereby
allowing the identity of the latter to be inferred. These
products may take the form of detoxified urinary metabolites of
the intermediate, or of adducts formed between the intermediate
and cellular macrostructures such as protein or nucleic acids.
Investigators are currently at work identifying the structures
of representative examples of each of these, and it may be
possible that prostaglandin synthase creates reactive intermediates (perhaps via electron withdrawal) that are uniquely
different from those which are produced by other enzymes (e.g.,
cytochrome P-450). If this possibility holds true, the unique
metabolites arising from the activity of prostaglandin synthase
may serve as suitable experimental observables to indicate the
involvement of the enzyme in whole cells or in vivo. As
described in section 2.2.3, the oxidation of 4-aminobiphenyl by
prostaglandin synthase results in preferential reaction of the

amine with cytosine, while the primary, cytochrome P-450-derived
N-hydroxylated intermediate of this amine ultimately reacted
preferentially with deoxyguanosine (86). Studies of the site of
4-aminobiphenyl addition to the DNA isolated from the bladder
lumen of dogs fed the compound could reveal a strong
preponderance for one over the other adduct. This could provide
a means of assessing the relative contribution of each enzyme
system. To further establish the relationship, the relative
amounts of each adduct could be altered by co-feeding known
inducers or inhibitors of one or the other activity. Two other
aromatic amine bladder carcinogens for which this distinction
appears to be feasible are 2-naphthylamine and benzidine (110).
With the former, unique naphthoquinoneimine intermediates are
participants in the prostaglandin synthase pathway and yield
nucleic acid adducts different from those derived from
N-hydroxynaphthylamine (a likely P-450 metabolite). The
oxidation of benzidine by the synthase yields a single DNA
adduct that is chromatographically distinct from the major
adduct found in liver, which was described in section 2.2.2 (68).

The detection of small, excretable molecular remnants of
electron-deficient cosubstrate oxidation products that were
detoxified before they covalently reacted with a macromolecule
could also provide a means of establishing the involvement of
the synthase in the metabolism of carcinogenic amines. One
class of such metabolites, which have a direct mechanistic
relationship to their reactive precursors, are the thioether
conjugates derived from glutathione (and possibly cysteine). It
is now recognized that a major function of intracellular
glutathione is the covalent addition to, and thus inactivation
of, such electrophiles (111). As described in a previous
section, thioether conjugates derived from the reactive
intermediates of acetaminophen (49) and benzidine (65) have
proven to be useful probes of the metabolic activation pathway.
For certain carcinogens, such as the aminothiazole-substituted
5-nitrofurans (section 2.2.4), prostaglandin synthase appears to
be the only enzyme capable of oxidative metabolism (66). If so,
thioether conjugates derived from this process could serve a
triple function. Elucidation of their structure could suggest

the identity of the reactive intermediate from which they were derived, and could also provide an important clue as to the manner in which covalent reaction with protein occurs, which is information that is difficult to obtain by other means. In addition, they could serve as detectable in vivo "markers" of metabolism by the synthase. It might be expected that the amounts of such markers would rise or fall, as would the corresponding nucleic acid adduct(s), under the influence of co-administered inducers or inhibitors of prostaglandin production. Once a firm relationship between marker excretion and synthase activity had been established, the levels of the markers excreted could in turn be used as an inexpensive means of screening combinations of inhibitor formulations without recourse to long term animal feeding studies and the eventual appearance (or lack thereof) of tumors. Only those inhibitor formulations that maximally depress synthase activity while showing minimal deleterious side effects would be utilized in a feeding regimen. The ultimate purpose of such studies would be the development of effective means of attenuating the malignant diseases that result from the interaction of carcinogens with prostaglandin synthase.

REFERENCES

1. Rehn L: Blasengeschwulste bei fuchsinarbeitern. Arch Klin Chir (50): 588-600, 1895.
2. Case RAM, Hosker ME, McDonald DB, Pearson JT: Tumours of the urinary bladder in workmen engaged in the manufacture and use of certain dyestuff intermediates in the British chemical industry. Br J Ind Med (11): 75-104, 1954.
3. Melick WF, Escue, HM, Naryka JJ, Mezera RA, Wheeler ER: The first reported cases of human bladder tumors due to a new carcinogen--xenylamine. J Urol (74): 760-766, 1955.
4. Deichmann WB, Radomski J, Glass E, Anderson, WAD, Coplan M, Woods F: Synergism among oral carcinogens. III. Simultaneous feeding of four bladder carcinogens to dogs. Ind Med Surg (34): 640-649, 1965.
5. Spitz S, Maguigan WH, Dobriner K: The carcinogenic action of benzidine. Cancer (3): 789-804, 1950.
6. Boyland E, Harris J, Horning, ES: The induction of carcinoma of the bladder in rats with acetamidofluorene. Br J Cancer (8): 647-654, 1954.
7. Marnett LJ: Polycyclic aromatic hydrocarbon oxidation during prostaglandin biosynthesis. Life Sci (29): 531-546, 1981.

8. Gale PH, Egan RW: Prostaglandin endoperoxide synthase catalyzed oxidation reactions. In: Pryor WA (ed) Free Radicals in Biology, Vol VI. Academic Press, Orlando, Florida, 1984, pp 1-38.

9. Marnett LJ, Eling TE: Cooxidation during prostaglandin biosynthesis: A pathway for the metabolic activation of xenobiotics. In: Hodgson E, Bend JB, Philpot RM (eds) Reviews in Biochemical Toxicology, Vol 5. Elsevier, New York, 1983, pp 135-172.

10. Sivarajah K, Lasker JM, Eling TE: Prostaglandin synthetase-dependent cooxidation of (+)-benzo(a)pyrene-7,8-dihydrodiol by human lung and other mammalian tissues. Cancer Res (41): 1834-1839, 1981.

11. Wise RW, Zenser TV, Kadlubar FF, Davis BB: Metabolic activation of carcinogenic aromatic amines by dog bladder and kidney prostaglandin H synthase. Cancer Res (44): 1893-1897, 1984.

12. Cohen SM, Zenser TV, Murasaki G, Fukushima S, Mattammal MB, Rapp NS, Davis BB: Aspirin inhibition of N-[4-(5-nitro-2-furyl)-2-thiazolyl]formamide-induced lesions of the urinary bladder correlated with inhibition of metabolism by bladder prostaglandin endoperoxide synthetase. Cancer Res (41): 3355-3359, 1981.

13. Van Der Ouderaa FJ, Buytenhek M, Nugteren DH, Van Dorp DA: Purification and characterisation of prostaglandin endoperoxide synthetase from sheep vesicular glands. Biochim Biophys Acta (487): 315-331, 1977.

14. Kulmacz RJ, Lands WEM: Prostaglandin H synthase: Stoichiometry of heme cofactor. J Biol Chem (259): 6358-6363, 1984.

15. Rollins TE, Smith WL: Subcellular localization of prostaglandin-forming cyclooxygenase in Swiss mouse 3T3 fibroblasts by electron microscopic immunocytochemistry. J Biol Chem (255): 4872-4875, 1980.

16. Roth GJ, Stanford N, Majerus PW: Acetylation of prostaglandin synthase by aspirin. Proc Nat Acad Sci USA (72): 3073-3076, 1975.

17. Walsh C: Enzymatic Reaction Mechanisms. W.H. Freeman and Co., San Francisco, 1979, pp 488-493.

18. Dophin D: The electronic configurations of catalases and peroxidases in their high oxidation states: A definitive assessment. Israel J Chem (21): 67-71, 1981.

19. Kalyanaraman B, Mason RP, Tainer B, Eling TE: The free radical formed during the hydroperoxide-mediated deactivation of ram seminal vesicles is hemoprotein-derived. J Biol Chem (257): 4764-4768, 1982.

20. McCarthy M-B, White RE: Functional differences between peroxidase compound I and the cytochrome P-450 reactive oxygen intermediate. J Biol Chem (258): 9153-9158, 1983.

21. White RE, Coon MJ: Oxygen activation by cytochrome P-450. Ann Rev Biochem (49): 315-356, 1980.

22. Rapp NS, Zenser TV, Brown WW, Davis BB: Metabolism of benzidine by a prostaglandin-mediated process in renal inner medullary slices. J Pharmacol Exp Ther (215): 401-406, 1980.

23. Egan RW, Gale PH, Baptista EM, Kennicott KL, VandenHeuvel WJA, Walker RW, Fagerness PE, Kuehl Jr. FA: Oxidation reactions by prostaglandin cyclooxygenase-hydroperoxidase. J Biol Chem (256): 7352-7361, 1981.
24. Zenser TV, Mattammal MB, Davis BB: Demonstration of separate pathways for the metabolism of organic compounds in rabbit kidney. J Pharmacol Exp Ther (208): 418-421, 1979.
25. Mohandas J, Duggin GG, Horvath JS, Tiller DJ: Metabolic oxidation of acetaminophen (paracetamol) mediated by cytochrome P-450 mixed-function oxidase and prostaglandin endoperoxide synthetase in rabbit kidney. Toxicol Appl Pharmacol (61): 252-259, 1981.
26. Boyd JA, Eling TE: Prostaglandin endoperoxide synthetase-dependent cooxidation of acetaminophen to intermediates which covalently bind in vitro to rabbit renal medullary microsomes. J Pharmacol Exp Ther (219): 659-664, 1981.
27. Egan RW, Gale PH, Kuehl Jr FA: Reduction of hydroperoxides in the prostaglandin biosynthetic pathway by a microsomal peroxidase. J Biol Chem (254): 3295-3302, 1979.
28. Egan RW, Gale PH, VandenHeuvel WJA, Baptista EM, Kuehl Jr FA: Mechanism of oxygen transfer by prostaglandin hydroperoxidase. J Biol Chem (255): 323-326, 1980.
29. Reed GA, Brooks EA, Eling TE: Phenylbutazone-dependent epoxidation of 7,8-dihydroxy-7,8-dihydrobenzo(a)pyrene. J Biol Chem (259): 5591-5595, 1984.
30. Siedlik PH, Marnett LJ: Oxidizing radical generation by prostaglandin H synthase. Methods Enzymol (105): 412-417, 1984.
31. Boyd JA, Harvan DJ, Eling TE: The oxidation of 2-aminofluorene by prostaglandin endoperoxide synthetase. J Biol Chem (258): 8246-8254, 1983.
32. Nelson RF: Anodic oxidation pathways of aliphatic and aromatic nitrogen functions. In: Weinberg N (ed) Techniques of electro-organic synthesis. Wiley, New York, 1974.
33. Ross SD, Finkelstein M, Rudd EJ: Anodic Oxidations. Academic Press, New York, 1975.
34. Adams RN: Electrochemistry at Solid Electrodes. Dekker, New York, 1969.
35. Flaks A, Flaks B: Induction of liver cell tumours in IF mice by paracetamol. Carcinogenesis (4): 363-368, 1983.
36. Hinson JA, Pohl LR, Monks TJ, Gillette JR: Minireview: Acetaminophen-induced hepatotoxicity. Life Sci (29): 107-116, 1981.
37. Volans GN: Self-poisoning and suicide due to paracetamol. Int Med Res (4): 7-13, 1976.
38. McMurtry RJ, Snodgrass WR, Mitchell JR: Renal necrosis, glutathione depletion, and covalent binding after acetaminophen. Toxicol Appl Pharmacol (46): 87-100, 1978.

39. Newton JF, Braselton Jr WE, Kuo C-H, Kluwe WM, Gemborys MW, Mudge GH, Hook JB: Metabolism of acetaminophen by the isolated perfused kidney. J Pharmacol Exp Ther (221): 76-79, 1982.
40. Armbrecht HJ, Birnbaum LS, Zenser TV, Mattammal MB, Davis BB: Renal cytochrome P-450's—electrophoretic and electron paramagnetic resonance studies. Arch Biochem Biophys (197): 277-284, 1979.
41. Saker BM, Kincaid-Smith P: Papillary necrosis in experimental analgesic nephropathy. Brit Med J (1): 161-162, 1969.
42. Josephy PD, Eling TE, Mason RP: Oxidation of p-aminophenol catalyzed by horseradish peroxidase and prostaglandin synthase. Mol Pharmacol (23): 461-466, 1983.
43. Andersson B, Larsson R, Rahimtula A, Moldeus P: Hydroperoxide-dependent activation of p-phenetidine catalyzed by prostaglandin synthase and other peroxidases. Biochem Pharmacol (32): 1045-1050, 1983.
44. Nelson SD, Dahlin DC, Rauckman EJ, Rosen GM: Peroxidase-mediated formation of reactive metabolites of acetaminophen. Mol Pharmacol (20): 195-199, 1981.
45. Moldeus P, Rahimtula A: Metabolism of paracetamol to a glutathione conjugate catalyzed by prostaglandin synthetase. Biochem Biophys Res Comm (96): 469-475, 1980.
46. Moldeus P, Andersson B, Rahimtula A, Berggren M: Prostaglandin synthetase catalyzed activation of paracetamol. Biochem Pharmacol (31): 1363-1368, 1982.
47. Hinson JA, Mitchell JR: N-hydroxylation of phenacetin by hamster liver microsomes. Drug Metab Dispos (4): 430-435, 1976.
48. Calder IC, Creek MJ, Williams PJ: N-hydroxyphenacetin as a precursor of 3-substituted 4-hydroxyacetanilide metabolites of phenacetin. Chem-Biol Interact (8): 87-90, 1974.
49. Miner DJ, Kissinger PT: Evidence for the involvement of N-acetyl-p-quinoneimine in acetaminophen metabolism. Biochem Pharmacol (28): 3285-3290, 1979.
50. Hinson JA, Pohl LR, Gillette JR: N-hydroxyacetaminophen: A microsomal metabolite of N-hydroxyphenacetain but apparently not of acetaminophen. Life Sci (24): 2133-2138, 1979.
51. Nelson SD, Forte AJ, Dahlin DC: Lack of evidence for N-hydroxyacetaminophen as a reactive metabolite of acetaminophen in vitro. Biochem Pharmacol (29): 1617-1620, 1980.
52. Rosen GM, Singletary Jr WV, Rauckman EJ, Killenberg PG: Acetaminophen hepatotoxicity: An alternative mechanism. Biochem Pharmacol (32): 2053-2059, 1983.
53. Rosen GM, Rauckman EJ, Ellington SP, Dahlin DC, Christie JL, Nelson SD: Reduction and glutathione conjugation reactions of N-acetyl-p-benzoquinone imine and two dimethylated analogues. Mol Pharmacol (25): 151-157, 1983.
54. Miller MG, Jollow DJ: Effect of L-ascorbic acid on acetaminophen-induced hepatotoxicity and covalent binding in hamsters. Drug Metab Dispos (12): 271-279, 1984.

55. Devalia JL, McLean AEM: Covalent binding and the
    mechanism of paracetamol toxicity. Biochem Pharmacol
    (32): 2602-2603, 1983.
56. de Vries J: Hepatotoxic metabolic activation of
    paracetamol and its derivatives phenacetin and benorilate:
    Oxygenation or electron transfer? Biochem Pharmacol
    (30): 399-402, 1981.
57. Porter KE, Dawson AG: Inhibition of respiration and
    gluconeogenesis by paracetamol in rat kidney
    preparations. Biochem Pharmacol (28): 3057-3062, 1979.
58. Trager WF: The postenzymatic chemistry of activated
    oxygen. Drug Metab Rev (13): 51-69, 1982.
59. Kadlubar FF, Miller, JA, Miller EC: Hepatic microsomal
    N-glucuronidation and nucleic acid binding of N-hydroxy
    arylamines in relation to urinary bladder carcinogenesis.
    Cancer Res (37): 805-814, 1977.
60. Radomski JL: The primary aromatic amines: Their
    biological properties and structure-activity relationships.
    Ann Rev Pharmacol Toxicol (19): 129-157, 1979.
61. Zenser TV, Mattammal MB, Davis BB: Demonstration of
    separate pathways for the metabolism of organic compounds
    in rabbit kidney. J Pharmacol Exp Ther (208): 418-421,
    1979.
62. Mattammal MB, Zenser TV, Brown WW, Herman CA, Davis BB:
    Mechanism of inhibition of renal prostaglandin production
    by acetaminophen. J Pharmacol Exp Ther (210): 405-409,
    1979.
63. Zenser TV, Mattammal MB, Davis BB: Cooxidation of
    benzidine by renal medullary prostaglandin cyclo-
    oxygenase. J Pharmacol Exp Ther (211): 460-464, 1979.
64. Zenser TV, Mattammal MB, Armbrecht HJ, Davis BB:
    Benzidine binding to nucleic acids mediated by the
    peroxidative activity of prostaglandin endoperoxide
    synthetase. Cancer Res (40): 2839-2845, 1980.
65. Rice JR, Kissinger PT: Cooxidation of benzidine by
    horseradish peroxidase and subsequent formation of
    possible thioether conjugates of benzidine. Biochem
    Biophys Res Comm (104): 1312-1318, 1982.
66. Wise RW, Zenser TV, Davis BB: Peroxidase metabolism of
    the urinary bladder carcinogen 2-amino-4-(5-nitro-2-furyl)-
    thiazole. Cancer Res (43): 1518-1522, 1983.
67. Josephy PD, Eling TE, Mason RP: An electron spin
    resonance study of the activation of benzidine by
    peroxidases. Mol Pharmacol (23): 766-770, 1982.
68. Martin CN, Beland FA, Roth RW, Kadlubar FF: Covalent
    binding of benzidine and N-acetylbenzidine to DNA at the
    C-8 atom of deoxyguanosine in vivo and in vitro. Cancer
    Res (42): 2678-2696, 1982.
69. Gemborys MW, Gribble GW, Mudge GH: Synthesis of
    N-hydroxyacetaminophen, a postulated toxic metabolite of
    acetaminophen, and its phenolic sulfate conjugate. J Med
    Chem (21): 649-652, 1978.
70. Miner DJ, Rice JR, Riggin RM, Kissinger PT: Voltammetry
    of acetaminophen and its metabolites. Anal Chem (53):
    2258-2263, 1981.

71. Rice JR, Kissinger PT: Determination of benzidine and its acetylated metabolites in urine by liquid chromatography. J Anal Toxicol (3): 64-66, 1979.
72. Zenser TV, Mattammal MB, Wise RW, Rice JR, Davis BB: Prostaglandin H synthase-catalyzed activation of benzidine: A model to assess pharmacologic intervention of the initiation of chemical carcinogenesis. J Pharmacol Exp Ther (227): 545-550, 1983.
73. Josephy PD, Eling T, Mason RP: The horseradish peroxidase-catalyzed oxidation of 3,5,3',5'-tetramethylbenzidine. J Biol Chem (257): 3669-3675, 1982.
74. Josephy PD, Eling TE, Mason RP: Co-oxidation of benzidine by prostaglandin synthase and comparison with the action of horseradish peroxidase. J Biol Chem (258): 5561-5569, 1983.
75. Claiborne A, Fridovich I: Chemical and enzymatic intermediates in the peroxidation of o-dianisidine by horseradish peroxidase. 1. Spectral properties of the products of dianisidine oxidation. Biochemistry (18): 2324-2328, 1979.
76. Wise RW, Zenser TV, Davis BB: Prostaglandin H synthase metabolism of the urinary bladder carcinogens benzidine and ANFT. Carcinogenesis (3): 285-289, 1983.
77. Frederick CB, Mays JB, Ziegler DM, Guengerich FP, Kadlubar FF: Cytochrome P-450- and flavin-containing monooxygenase-catalyzed formation of the carcinogen N-hydroxy-2-aminofluorene and its covalent binding to nuclear DNA. Cancer Res (42): 2671-2677, 1982.
78. Saunders BC, Holmes-Siedle AG, Stark BP: Peroxidase. Washington, Butterworths, 1964, p 9.
79. Josephy PD, Mason RP, Eling T: Chemical structure of the adducts formed by the oxidation of benzidine in the presence of phenols. Carcinogenesis (3): 1227-1230, 1982.
80. Josephy PD, Damme AV: Reaction of 4-substituted phenols with benzidine in a peroxidase system. Biochem Pharmacol (33): 1155-1156, 1984.
81. McKee RH, Tometsko AM: Inhibition of promutagen activation by the antioxidants butylated hydroxyanisole and butylated hydroxytoluene. J Natl Cancer Inst (63): 473-477, 1979.
82. Maeura Y, Weisburger JH, Williams GH: Dose-dependent reduction of N-2-fluorenylacetamide-induced liver cancer and enhancement of bladder cancer in rats by butylated hydroxytoluene. Cancer Res (44): 1604-1610, 1984.
83. Weber WW: Acetylation pharmacogenetics: Experimental models for human toxicity. Fed Proc (43): 2332-2337, 1984.
84. Lower Jr GM, Nilsson T, Nelson CE, Wolf H, Gamsky TE, Bryan GT: N-Acetyltransferase phenotype and risk in urinary bladder cancer: Approaches in molecular epidemiology. Preliminary results in Sweden and Denmark. Environ Health Perspect (29): 71-79, 1979.
85. Lower Jr GM, Bryan GT: Enzymatic N-acetylation of carcinogenic aromatic amines by liver cytosol of species displaying different organ susceptibilities. Biochem Pharmacol (22): 1581-1588, 1973.

86. Morton KC, King CM, Vaught JB, Wang CY, Lee M-S, Marnett LJ: Prostaglandin H synthase-mediated reaction of carcinogenic arylamines with tRNA and homopolyribonucleotides. Biochem Biophys Res Comm (111): 96-103, 1983.

87. Kadlubar FF, Frederick CB, Weis CC, Zenser TV: Prostaglandin endoperoxide synthetase-mediated metabolism of carcinogenic aromatic amines and their binding to DNA and protein. Biochem Biophys Res Comm (108): 253-258, 1982.

88. Radomski JL, Rey AA, Brill E: Evidence for a glucuronic acid conjugate of N-hydroxy-4-aminobiphenyl in the urine of dogs given 4-aminobiphenyl. Cancer Res (33): 1284-1289, 1973.

89. Radomski JL, Hearn WL, Randomski T, Moreno H, Scott WE: Isolation of the glucuronic acid conjugate of N-hydroxy-4-aminobiphenyl from dog urine and its mutagenic activity. Cancer Res (37): 1757-1762, 1977.

90. Miller JA: Carcinogenesis by chemicals: An overview—G.H.A. Clowes Memorial Lecture. Cancer Res (30): 559-576, 1970.

91. Vaught JB, Lee M-S, Shayman MA, Thissen MR, King CM: Arylhydroxylamine-induced ribonucleic acid chain cleavage and chromatographic analysis of arylamine-ribonucleic acid adducts. Chem-Biol Interact (34): 109-124, 1981.

92. Nitrofurans: Chemistry, Metabolism, Mutagenesis, and Carcinogenesis. Bryan GT (ed) Carcinogenesis--A comprehensive survey, Vol. 4. New York, Raven Press, 1978.

93. Jacobs JB, Arai M, Cohen SM, Friedell GH: A long-term study of reversible and progressive urinary bladder lesions in rats fed N-[4-(5-nitro-2-furyl)-2-thiazolyl] formamide. Cancer Res (37): 2817-2821, 1977.

94. Cohen SM, Lower Jr GM, Erturk F, Bryan GT: Comparative carcinogenicity in Swiss mice of N-[4-(5-nitro-2-furyl)-2-thiazolyl]acetamide and structurally related 5-nitrofurans and 4-nitrobenzenes. Cancer Res (33): 1593-1597, 1973.

95. Wang CY, Kamiryo Y, Croft WA: Carcinogenicity of 2-amino-4-(5-nitro-2-furyl)thiazole in rats by oral and subcutaneous administration. Carcinogenesis (3): 275-277, 1982.

96. Erturk E, Cohen SM, Bryan GT: Carcinogenicity of N-[4-(5-nitro-2-furyl)-2-thiazolyl]acetamide in female rats. Cancer Res (30): 936-941, 1970.

97. Croft WA, Bryan GT: Production of urinary bladder carcinomas in male hamsters by N-[4-(5-nitro-2-furyl)-2-thiazolyl]formamide, N-[4-(5-nitro-2-furyl)-2-thiazolyl] acetamide, or formic acid 2-[4-(5-nitro-2-furyl)-2-thiazolyl]hydrazide. J Natl Cancer Inst (51): 941-949, 1973.

98. Wang CY, Behrens BC, Ichikawa M, Bryan GT: Nitroreduction of 5-nitrofuran derivatives by rat liver xanthine oxidase and reduced nicotinamide adenine dinucleotide phosphate-cytochrome c reductase. Biochem Pharmacol (23): 3395-3404, 1974.

99. Mattammal MB, Zenser TV, Davis BB: Anaerobic metabolism and nuclear binding of the carcinogen 2-amino-4-(5-nitro-2-furyl)thiazole (ANFT). Carcinogenesis (3): 1339-1344, 1982.

100. Swaminathan S, Bryan GT: Biotransformation of the bladder carcinogen N-[4-(5-nitro-2-furyl)-2-thiazolyl] formamide in mice. Cancer Res (44): 2331-2338, 1984.

101. Wang CY, Hayashida S, Pamukcu AM, Bryan GT: Enhancing effect of allopurinol on the induction of bladder cancer in rats by N-[4-(5-nitro-2-furyl)-2-thiazolyl]formamide (FANFT). Proc Am Assoc Cancer Res (18): 100, 1976.

102. Zenser TV, Palmier MO, Mattammal MB, Bolla RI, Davis BB: Comparative effects of prostaglandin H synthase-catalyzed binding of two 5-nitrofuran urinary bladder carcinogens. J Pharmacol Exp Ther (227): 139-143, 1983.

103. Wang CY, Bryan GT: Deacylation of carcinogenic 5-nitrofuran derivatives by mammalian tissues. Chem-Biol Interact (9): 423-428, 1974.

104. Zenser TV, Palmier MO, Mattammal MB, Davis BB: Metabolic activation of the carcinogen N-[4-(5-nitro-2-furyl)-2-thiazolyl]acetamide by prostaglandin H synthase. Carcinogenesis, in press.

105. Rice JR, Kissinger PT: Liquid chromatography with precolumn sample preconcentration and electrochemical detection: Determination of aromatic amines in environmental samples. Environ Sci Technol (16): 263-268, 1982.

106. Zenser TV, Mattammal MB, Rapp NS, Davis BB: Effect of aspirin on metabolism of acetaminophen and benzidine by renal inner medulla prostaglandin hydroperoxidase. J Lab Clin Med (101): 58-65, 1983.

107. Murasaki G, Zenser TV, Davis BB, Cohen SM: Inhibition by aspirin of N-[4-(5-nitro-2-furyl)-2-thiazolyl]formamide-induced bladder carcinogenesis and enhancement of fore-stomach carcinogenesis. Carcinogenesis (5): 53-55, 1984.

108. Cramer JW, Miller JA, Miller EC: N-hydroxylation: A new metabolic reaction observed in the rat with carcinogen 2-acetylaminofluorene. J Biol Chem (235): 885-888, 1960.

109. Miller EC, Miller JA: Searches for ultimate chemical carcinogens and their reactions with cellular macromolecules. Cancer (47): 2327-2345, 1981.

110. Yamazoe Y, Miller DW, Gupta RC, Zenser TV, Weis CC, Kadlubar FF: DNA adducts formed by prostaglandin H synthase-mediated activation of carcinogenic arylamines. Proc Am Assoc Cancer Res (25): 91, 1984.

111. Tateiski M: Methylthiolated metabolites. Drug Metab Rev (14): 1207-1234, 1983.

ACKNOWLEDGEMENTS

This work was supported by the Veterans Administration, U.S. Public Health Service Grant CA-28015 from the National Cancer Institute through the National Bladder Cancer Project, and the American Cancer Society, Missouri Chapter. The authors wish to thank Sharon Smith for skillful assistance in the preparation of this manuscript.

5

INVOLVEMENT OF PROSTAGLANDIN SYNTHASE IN THE METABOLIC ACTIVA-
TION OF ACETAMINOPHEN AND PHENACETIN

PETER MOLDÉUS, ROGER LARSSON AND DAVID ROSS

1. INTRODUCTION

Acetaminophen (paracetamol, N-acetyl-p-aminophenol) and
phenacetin (4-ethoxyacetamide) are analgesic and antipyretic
drugs. Acetaminophen is widely used whereas phenacetin has
been withdrawn from the market in most countries.

Acetaminophen is generally considered non-toxic and safe
when used at therapeutic doses. However, when acetaminophen
is taken in very high doses severe hepatic necrosis, often
associated with renal damage occurs (1-5). It is currently
believed that the microsomal cytochrome P-450 linked mono-
oxygenase system is responsible for activating acetaminophen
in the liver to an electrophilic intermediate that can bind
covalently to cellular macromolecules to produce cell damage
(6,7). In the presence of reduced glutathione (GSH) the reactive species is trapped as the corresponding glutathione conjugate (8,9). The acute kidney damage associated with the liver damage may depend on cytochrome P-450 dependent activation either in the liver or directly in the kidney (10).
There is also evidence of renal damage occurring without significant hepatic damage, both acute tubular necrosis at therapeutic doses of acetaminophen as well as analgesic nephropathy in patients chronically consuming analgesics containing acetaminophen (11-15).

Phenacetin is substantially more nephrotoxic than acetaminophen. There are several reports of different kinds of se-
rious kidney damage in man following prolonged intake of phenacetin including renal papillary necrosis (16,17). There

are also indications of an increased rate of cancer in the
kidney and urinary tract after long-term phenacetin exposure (18,19). Phenacetin has also been shown to induce tumours
of the renal pelvis and the lower urinary tract of rats (20,
21).

The metabolism of phenacetin in vivo is quite complex
and involves many different pathways. The major primary metabolite of phenacetin in man and experimental animals is
acetaminophen (22), which is excreted mainly as the sulfate
or glucuronide conjugate. Deacetylation to p-phenetidine is
also a major metabolic pathway (22) whilst other minor metabolites are 2-hydroxyphenacetin (22,23), p-aminophenol and
N-hydroxyphenacetin.

Acetaminophen has been suggested to be responsible for
the therapeutic effects of phenacetin but it seems less likely that it also causes the nephrotoxicity seen after phenacetin abuse since acetaminophen itself is less toxic than
phenacetin. The nephrotoxicity of phenacetin has been suggested to be caused by one of the minor metabolites, N-
hydroxyphenacetin (24,25,26). This metabolite may be conjugated with sulfate, glucuronic acid or acetate in the liver,
transported to the kidney where the conjugates could undergo hydrolysis to a reactive intermediate (27). In support
of this hypothesis is the finding that acetylation and sulfate conjugation activate N-hydroxyphenacetin to mutagenic
and nucleic acid binding metabolites (28,29). p-Aminophenol
another minor metabolite of phenacetin is a known nephrotoxic agent which may also be partly responsible for the
toxic effect of phenacetin. p-Aminophenol has also been implicated to be responsible for the nephrotoxicity of acetaminophen and was recently shown to be formed from acetaminophen in the kidney (30,31).

Cytochrome P-450 is present in the kidney and is located
almost exclusively in the renal cortex (32). The activity
is low compared to the liver but cytochrome P-450 dependent
oxidation of acetaminophen may be involved in the kidney
damage observed experimentally in animals (10) and also

173

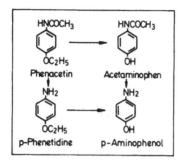

FIGURE 1.   Routes of metabolism of phenacetin.

after a massive overdose of acetaminophen in man.

Prostaglandin synthase (PGS) is present in the kidney and a particularly high activity is present in the inner medulla (33,34). PGS catalyzes the oxygenation of polyunsaturated fatty acids to hydroxy endoperoxides (e.g.PGH$_2$) (35). The most important substrate _in vivo_ is arachidonic acid (AA). PGS contains two activities. The fatty acid cyclooxygenase activity catalyzes the oxygenation of AA to a hydroperoxy endoperoxide (PGG$_2$), while the hydroperoxidase activity catalyses the reduction of PGG$_2$ to the hydroxy endoperoxide (PGH$_2$) (36,37). PGH$_2$ represents a branching point in the metabolism of AA and it can undergo tissue specific metabolism to different biological active substrates. PGS is associated with the endoplasmic reticulum and nuclear membranes, and is thus present in microsomal preparations. The cyclooxygenase activity of PGS is inhibited by nonsteroidal anti-inflammatory agents (38) such as indomethacin and acetylsalicylic acid, while there are no known specific inhibitors of the hydroperoxidase activity.

During the last few years an increasing number of chemicals many of them carcinogens, have been shown to undergo biotransformation during PG synthesis. This biotransformation, which often results in the formation of reactive metabolites, is catalyzed by the hydroperoxidase activity of the enzyme and occurs through a cooxidation mechanism as described in several recent reviews dealing with PGS catalyzed metabolism of xenobiotics (39-41).

We have in our laboratory been particularly interested in the involvement of PGS in the metabolism and metabolic activation of acetaminophen and phenacetin and its significance for the renal toxicity of these drugs. This manuscript attempts to summarize our own and others findings regarding the PGS catalyzed metabolism of acetaminophen, phenacetin and the phenacetin metabolites p-phenetidine and p-aminophenol to reactive products. The significance of these findings with regard to the nephrotoxicity of acetaminophen and phenacetin is also discussed.

2. ACETAMINOPHEN

PGS was first shown in 1980 to catalyze the formation of an acetaminophen glutathione conjugate (42). This study was performed using ram seminal vesicle microsomes (RSVM), a commonly used source of PGS, and the finding reflected a formation of a reactive intermediate which then conjugated with GSH. In subsequent studies irreversible binding to protein of acetaminophen during prostaglandin synthesis using RSVM, rabbit kidney medulla microsomes and purified PGS was reported (43-45).

The PGS dependent metabolic activation of acetaminophen was shown to be enzymatic, dependent on arachidonic acid and oxygen and inhibited by cyclooxygenase inhibitors such as indomethacin or antioxidants such as butylated hydroxyanisole (BHA) (Table 1). These results clearly indicated that PGS catalyzed the oxidation of acetaminophen and this was shown unequivocally when purified PGS was shown to catalyze acetaminophen metabolism (44). In agreement with results using other substrates it is evidently the peroxidase activity of PGS catalyzing the reduction of PGG_2 to PGH_2, which is responsible for the metabolic activation of acetaminophen. This conclusion is supported by the observation that AA can be replaced by linolenic acid hydroperoxide which partakes in a reaction not dependent on oxygen or inhibited by indomethacin (Table 1; 45). Linolenic acid hydroperoxide has

Table 1. Arachidonic acid and linolenic acid hydroperoxide dependent acetaminophen-glutathione conjugate formation in ram seminal vesicle microsomes.

Incubation conditions	Glutathione conjugates nmol/mg protein/min	
	+Arachidonic acid (100 µM)	+Linolenic acid hydroperoxide (100 µM)
Control[1]	27.8	15.0
Boiled microsomes	1.3	0.8
N$_2$ Atmosphere	5.9	14.7
Indomethacin, 100 µM	5.7	16.3
BHA[2], 500 µM	N.D.[3]	2.8

[1] Incubations were run for 1 min at 25°C and contained acetaminophen 200 µM; GSH 2.5 mM and protein 1 mg/ml in 100 mM phosphate buffer, pH 8.0.

[2] BHA = Butylated hydroxy anisole.

[3] N.D. = Not detected.

also been demonstrated to be effective in supporting the metabolism of acetaminophen to protein binding products (44). Thus, acetaminophen serves as a hydrogen donor for the peroxidase and cooxidation occurs during the synthesis of prostaglandins from AA. Furthermore acetaminophen increases the PG synthesis measured as oxygen uptake from AA (Fig. 2; 44,45). This has previously been demonstrated with phenol (46) and is probably mediated by the same mechanism with acetaminophen. Acetaminophen has also been shown to be a weak inhibitor of the cyclooxygenase activity of PGS (47). This inhibition however is not evident with RSVM until concentrations of acetaminophen greater than 1 mM are reached (Fig. 2).

The affinity of acetaminophen for the PGS catalyzed activation is very high and maximal activity is reached at 200 µM (45). This is actually a much higher affinity than for the cytochrome P-450 dependent reaction. The K$_m$ for the PGS dependent reaction is about 40 µM, which is less than one tenth of that of the cytochrome P-450 catalyzed reaction (Table 2). The high affinity for PGS may have some important

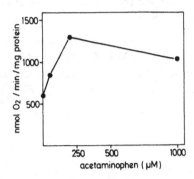

FIGURE 2. Effect of acetaminophen on the oxygenation of arachidonic acid by RSV microsomes.

implications. Whereas the cytochrome P-450 dependent activation of acetaminophen might only be functioning at full capacity after a massive overdose when the plasma concentrations of the drug reach millimolar concentrations, the PGS reaction may be functioning at therapeutic doses of acetaminophen. The activity of the PGS reaction is also very much higher (Table 2; 45) but the reaction only lasts for a short while due to the self-deactivation of the enzyme (48).

Table 2. K_m and V_{max} for acetaminophen-glutathione conjugate formation in microsomes from RSV and mouse liver.

Microsome source	K_m (µM)	V_{max} (nmoles/min per mg protein)
RSV[1]	54	100
Mouse liver[2]	429	0.75

[1] In the presence of arachidonic acid (0.3 mM)
[2] In the presence of a NADPH generating system

The mechanism of activation of acetaminophen by PGS is not yet established, nor is the nature of the reactive metabolite. The reactive metabolite binding to GSH seems to be identical to that formed during cytochrome P-450 dependent oxidation of acetaminophen. This metabolite has been suggested to be N-acetyl-p-benzoquinone imine (NAPQI) (49-51). The formation of such a reactive intermediate may proceed by two one electron steps or directly by a two electron pathway. For cytochrome P-450 the one electron pathway would involve a reaction of acetaminophen with a ferryl oxy radical complex of cytochrome P-450, resulting in a hydrogen abstraction to form a semiquinone or nitrenium radical. This acetaminophen radical could then easily be oxidized by a rapid second electron transfer to produce NAPQI and a hydrated ferric cytochrome P-450 complex (52). Alternatively NAPQI may be formed directly. In this case a perferryl form of cytochrome P-450 would react with acetaminophen to give a ferric oxyamide complex which could readily decompose to give NAPQI.

Peroxidase-catalyzed reactions such as that catalyzed by PGS are generally considered to occur via the production of a substrate-derived free radical and a one electron oxidized form of the enzyme. This would imply that the PGS dependent metabolic activation of acetaminophen proceeds via a one electron pathway with the intermediary formation of an acetaminophen free radical presumably the phenoxy radical. That such a radical exists is supported by a number of findings. Even though no radical intermediate has so far been detected directly by ESR during the PGS catalyzed reaction such a radical has been identified in a similar reaction using horse radish peroxidase and hydrogen peroxide (HRP/ H_2O_2) (52). Furthermore GSH, if present during PGS catalyzed metabolism of acetaminophen, not only forms conjugates with a reactive species but is also extensively oxidized to GSSG (Fig. 3; 45). This oxidation of GSH can be ascribed to the interaction of a radical intermediate of acetaminophen (R˙) resulting in the formation of a thiyl radical (GS˙) and regeneration of acetaminophen; GSSG formation can subsequently occur by dimerization of two GS˙ radicals.

$$R˙ + GSH \longrightarrow RH + GS˙$$
$$GS˙ + GS˙ \longrightarrow GSSG$$

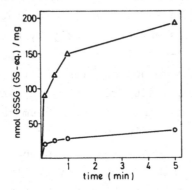

FIGURE 3. Time-course of oxidation of GSH by RSV microsomes in the presence of AA (0.2 mM) (o) or AA + acetaminophen (200 μM) (Δ). The initial GSH concentration was 1 mM.

Other radicals such as those derived from chlorpromazine have been shown to be reduced by GSH in a similar manner (53).

If an acetaminophen radical intermediate is formed during the PGS catalyzed reaction it may be further oxidized, either by PGS or nonenzymatically, to NAPQI prior to reacting with GSH or tissue protein. The radical itself may also react with GSH to form a glutathione conjugate, a reaction which has recently been proposed (54). Identical glutathione conjugates are formed by PGS and cytochrome P-450 dependent oxidation of acetaminophen indicating that the same reactive species, presumably NAPQI, is formed in both reactions (Fig. 4; 45).

The PGS catalyzed metabolic activation of acetaminophen is of particular interest regarding the nephrotoxicity of the drug. Acetaminophen may cause two types of nephrotoxicity. After a massive overdose, in addition to the hepatic lesion, acute tubular necrosis may be produced. This is primarily a lesion of the renal cortex and seem to depend on cytochrome P-450 dependent activation in this tissue. With chronic abuse of acetaminophen however, the initial lesion is localized primarily to the medulla. It is this latter type of lesion which could result from a PGS-catalyzed acti-

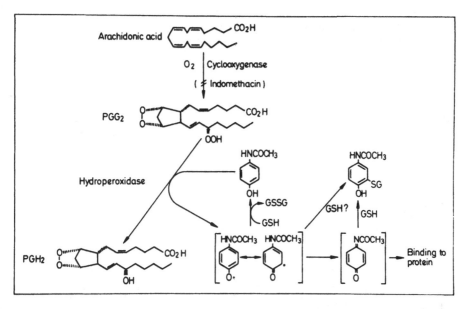

FIGURE 4. Possible mechanism for the PGS-catalyzed metabolic activation of acetaminophen.

vation of acetaminophen. The PGS activity in the renal inner
medulla is very high whereas cytochrome P-450 activity is
virtually absent (32). Furthermore acetaminophen has been
shown to reach higher concentration in the cells of the renal
inner medulla relative to the cortex (55). In this context
the high affinity of PGS for acetaminophen may be an important determinant of acetaminophen nephrotoxicity during chronic abuse.

That PGS of the inner medulla does catalyze the metabolism of acetaminophen to reactive protein binding metabolites has been shown in three studies (Table 3). The activity,
which varied between the different studies, was considerably
lower than in RSVM, but was considerably higher however than
the corresponding NADPH-dependent binding in this tissue.

Table 3. Arachidonic acid-dependent irreversible binding of acetaminophen to rabbit renal medullary microsomes.

Acetaminophen µM	Arachidonic acid µM	Acetaminophen bound pmol/10 min per mg protein	Ref.
200	100	454	45
100	200	1028	44
500	25	1285 (109)[1]	43

[1]The value in brackets represents NADPH supported activity

3. PHENACETIN

Phenacetin is metabolized in vivo to a number of primary metabolites. Quantitatively the most important of these are acetaminophen and p-phenetidine. Phenacetin itself is not a substrate for PGS but some of its metabolites in addition to acetaminophen are (56).

The PGS dependent metabolism of the two metabolites, p-phenetidine and p-aminophenol was shown to result in the production of genotoxic species which induced single strand breaks in human fibroblast DNA. This was demonstrated in experiments where cultured human fibroblasts were coincubated with RSVM, AA and a metabolite (Table 4; 56). The activation to genotoxic products was dependent on AA and inhibited by indomethacin as well as by acetylsalicylic acid which clearly indicates the involvement of PGS. Both p-phenetidine and p-aminophenol had very high affinity for the PGS reaction and a substrate concentration of 50 µM produced a maximal response (56). Acetaminophen did not induce single strand breaks in this system, eventhough it was metabolized to a reactive product, capable of interaction with protein and glutathione, by PGS. Evidently NAPQI, assuming this is the ultimate reactive product, does not interact with DNA.

p-Aminophenol has long been recognized to be nephrotoxic and has recently been shown to be metabolized through to a radical intermediate by both PGS and HRP (55). p-Aminophenol is however probably only a very minor metabolite of phe-

Table 4. Effect of phenacetin and phenacetin metabolites on the induction of DNA strand breaks in human fibroblasts.

Incubation conditions	Fraction of single stranded DNA %
Solvent control (DMSO)	24.3±7.3
Phenacetin (200 µM)	20.5±7.5
Acetaminophen (200 µM)	21.6±5.2
p-Aminophenol (50 µM)	39.3±8.2
p-Phenetidine (50 µM)	41.5±9.8
p-Phenetidine - Arachidonic acid	20.3±4.5
p-Phenetidine + Indomethacin, 100 µM	20.8±2.8
p-Phenetidine + Acetylsalicylic acid, 1 mM	24.7±2.3

Incubations were performed as described in ref. 56. Arachidonic acid concentration was 100 µM and RSV microsomal protein 1 mg/ml.

nacetin and its production would therefore seem to be of limited importance to the nephrotoxicity associated with phenacetin.

Since p-phenetidine is a major metabolite of phenacetin it is thus potentially of greater importance for any putative PGS-dependent mechanism of phenacetin nephrotoxicity, we decided to investigate the metabolism of this particular phenacetin metabolite in more detail. Our aim was to elucidate the mechanism of the PGS dependent metabolism of p-phenetidine and to identify the products, formed in this reaction particularly those which would react with biological nucleophiles.

PGS dependent metabolism had so far been shown to result in DNA damaging product(s). The metabolism was subsequently demonstrated to result in protein binding as well as DNA binding (Table 5; 58,59). The protein binding was extensive whereas the binding to DNA was low but significant. As expected the AA supported binding to both protein and DNA was inhibited by indomethacin and the antioxidant BHA. As was the case with acetaminophen, AA could be replaced by linolenic acid hydroperoxide indicating that the peroxidase activity of PGS was responsible. In addition to organic hydro-

Table 5. Prostaglandin synthase dependent binding of (^{14}C)-
p-phenetidine to protein and DNA.

Incubations conditions	p-phenetidine bound	
	nmol/mg protein /30 sec	pmol/mg DNA/min
Complete system	15 2±1.2	21±5
Boiled microsomes	0	<5
- Arachidonic acid	0	<5
+ Indomethacin, 100 µM	0.5±0.2	<5
+ GSH, 1 mM	2.9±0.6	71±9
- Arachidonic acid + Hydrogen peroxide, 100 µM	16.6±0.2	17±1

Complete system: RSV microsomes (1 mg/ml); p-phenetidine 50
µM; EDTA 1 mM; phosphate buffer 100 mM, pH 8.0 and arachidonic acid 100 µM. Reactions were started by the addition of
arachidonic acid.

peroxides, H_2O_2 was also able to support the PGS dependent
reaction with p-phenetidine (Table 5). This may be of physiological relevance since H_2O_2 is a product of several cellular enzymatic reactions eventhough the hydroperoxidase of
PGS would presumably have to compete with cellular catalase
and glutathione peroxidase for the H_2O_2 available. Furthermore H_2O_2 is generated by leucocytes in glomerular capillar-
ies (60) and have also been found to stimulate phospholipase
activity in isolated glomeruli and thus increase the availability of AA (1). Horseradish peroxidase was also shown to
metabolize p-phenetidine to protein- and DNA binding prod-
uct(s) (58). This is thus a further indication that it is
the hydroperoxidase of PGS that catalyzes the metabolism of
p-phenetidine.

Thiols such as GSH or other nucleophiles usually inhibit
protein and DNA binding by either conjugating or reducing
the reactive species. This is true also for the peroxidase
catalyzed protein binding of p-phenetidine and almost no
binding was observed in the presence of 1 mM GSH (Table 5;
58). Surprisingly however we found that the binding to DNA
was actually enhanced several fold in the presence of GSH
(Table 5; 59). This binding was presumably caused by one or
more GSH conjugates formed in the presence of GSH but was

not covalent since most of it could be removed by dialysis
(59). The physiological relevance of this association of
glutathione conjugates with DNA is unclear.

Peroxidase catalyzed reactions, such as the PGS reactions, are generally considered to occur via the production
of substrate derived free radicals and a one electron oxidized form of the peroxidase enzyme. Despite this the formation of substrate-derived free radicals has only been demonstrated in a few cases, namely acetaminophen (52), aminopyrine (61), 1-phenyl-3-pyrazolidone (62), benzidine (63),
3,5,3',5'-tetramethylbenzidine (64,65) and p-aminophenol
(57). The peroxidase catalyzed metabolism of p-phenetidine
probably also proceeds via the initial formation of a p-
phenetidine free radical. We have been unable to demonstrate the presence of such a radical intermediate using ESR
spectroscopy. However, other observations indicate the presence of a p-phenetidine free radical in the peroxidase-
catalyzed metabolism of p-phenetidine. For instance inclu-
sion of GSH in a PGS/AA reaction with p-phenetidine results
in a decreased consumption of p-phenetidine which is not
accompanied by a decrease in the utilization of AA (66).
GSH is rapidly oxidized in the reaction and this implies
that GSH can serve as an electron donor to a p-phenetidine
radical resulting in the formation of a glutathione thiyl
radical which subsequently dimerizes to form GSSG (56,66).
This is in agreement with results obtained from investigations of the PGS dependent oxidation of acetaminophen (45).

In a peroxidase-mediated reaction the enzyme undergoes
a divalent oxidation by a peroxide co-factor. The reduction
of this oxidized enzyme (Compound I) may occur by a divalent mechanism or by two successive univalent reductions.
The latter mechanism results in the generation of substrate-derived free radicals and if this is the sole route of
reduction a ratio of substrate consumed to peroxide cofactor used of 2/1 would be expected. In experiments using the
HRP/H_2O_2 system we observed that the ratio between p-phene-

tidine consumed and H_2O_2 used was 2:1 which strongly supports a mechanism which involves two successive one electron
reductions of an oxidized enzyme complex (66). By definition
then this mechanism would involve the generation of substrate-derived free radicals.

Additional evidence for a p-phenetidine radical intermediate emerged from studies of oxygen uptake in the HRP/H_2O_2
catalyzed oxidation of p-phenetidine. The presence of thiols
in incubations containing HRP/H_2O_2 and p-phenetidine caused
extensive oxygen uptake (66). Similar behaviour has been observed with aminopyrine (67) and acetaminophen (69), both
of which have been shown to form free radical intermediates
in peroxidase-catalyzed reactions by ESR spectroscopy. The
p-phenetidine radical itself does not combine directly with
oxygen as in the absence of thiols no oxygen uptake occurred.
These results suggest therefore the interaction of a thiyl
radical with molecular oxygen. Such reactions of glutathio-
ne-derived thiyl radicals are well documented (69,70). They
give rise to unidentified oxygen-containing products which
have been suggested to be sulfinic and sulfonic acid derivatives of the thiol, which are formed according to equations
1 $\longrightarrow$ 2 (71).

Equation 1/ $GS^{\cdot} + O_2 \longrightarrow GSO_2^{\cdot}$

Equation 2/ $2GSO_2^{\cdot} + H_2O \longrightarrow GSO_2H + GSO_3H$

The PGS dependent oxidation of p-phenetidine resulted in
the formation of several intensely coloured products which
could be separated by thin layer chromatography (TLC) (Table 6; 66). The broad spectrum of coloured products formed
in the peroxidase catalyzed oxidation of p-phenetidine
lends further support to a radical mechanism. The formation
of coloured oligomers in the peroxidase-catalyzed oxidation
of aromatic amines and phenols due to radical coupling reactions is well established (72), and is common when primary
aromatic amines are subjected to anodic oxidation (73). A
very similar metabolite pattern was produced if HRP and

Table 6. Characteristics of p-phenetidine metabolites formed in prostaglandin synthase and horseradish peroxidase catalyzed reactions.

Metabolite	RSVM/AA		HRP/H_2O_2	
	R_f	Colour	R_f	Colour
7	0.74	Pink	-	-
6	0.70	Yellow	0.70	Yellow
5	0.65	Orange	0.65	Orange
4	0.56	Purple	0.56	Purple
3	0.25	Brown	0.25	Brown
2	0.10	Red	0.10	Red
1	0	Red	0	Red

Incubations and thin layer chromatography were performed as described in ref. 66.

H_2O_2 was used in place of PGS and AA suggesting a common mechanism of oxidation of p-phenetidine in the two systems. A similar mechanism for the oxidation of benzidine in the two systems has also recently been proposed (63). The HRP/ H_2O_2 system was therefore used for mechanistic studies on the peroxidase catalyzed metabolism of p-phenetidine as well as to generate the various products for isolation and identification.

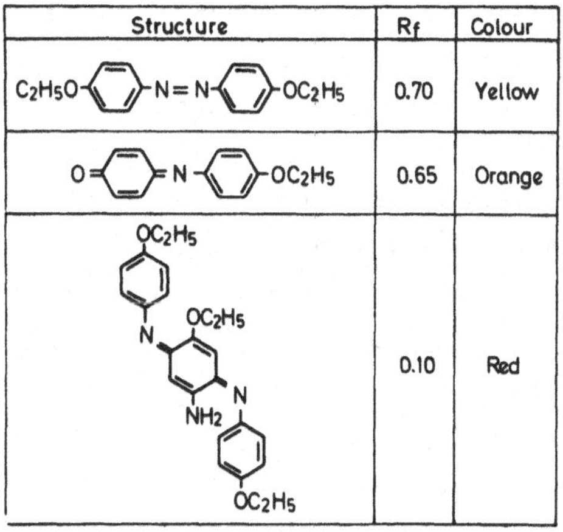

FIGURE 5. Chemical structures of p-phenetidine metabolites formed in peroxidase reaction.

The composition and identity of three of the major coloured bands produced during peroxidase catalyzed oxidation
of p-phenetidine and isolated by TLC was established by
mass spectral analysis (66,74) (see Fig. 5). The yellow
band (R_f 0.70) contained a di-azo compound. This is particularly relevant for mechanistic reasons as the formation
of di-azo compounds would be expected if the oxidation of
p-phenetidine proceeded via the production of nitrogen-
centered free radicals. The mass spectrum of the red metabolite (R_f 0.10) indicated a p-phenetidine trimer and NMR studies of this compound were consistent with this structure.

The orange metabolite was identified as a quinone-imine
dimer of p-phenetidine (74). As indicated by its structure
this metabolite was electrophilic and bound extensively to
added protein. It also reacted with GSH and at least two
conjugates were formed, one of which was probably a di-glu-
tathione conjugate (75). The orange compound could also be
readily reduced to the colourless hydroquinone either non-
enzymatically with dithionite, ascorbic acid or dithiothrei-
tol or enzymatically by NADPH cytochrome P-450 reductase
and as expected this reduced form did not react with protein or GSH. The orange quinone-imine dimer of p-phenetidine
did not bind to DNA or induce single strand breaks in fibroblast DNA and could thus not be responsible for the genotoxic effects seen during the PGS catalyzed oxidation of
p-phenetidine.

It could however be responsible for the protein binding
observed during p-phenetidine metabolism. That the major
protein binding species produced during peroxidase catalyzed oxidation of p-phenetidine was a dimeric species was
confirmed by the difference in protein binding between ring
(^{14}C) and ethyl (^{14}C) p-phenetidine (74). The ratio of protein binding of ring label to ethyl label of approximately
2:1 is consistent with the binding species beeing a p-phenetidine dimer. This would necessitate the loss of an ethoxy
group from one of the two p-phenetidine molecules involved
in the formation of the dimer. Ethanol was indeed isolated

as a product of the peroxidase catalyzed oxidation of p-phenetidine which supported this mechanism (74).

It was difficult to understand how a p-phenetidine qui-none-imine dimer could be formed directly from radical re-actions. An alternative mechanism for the formation of the quinone-imine dimer could involve the initial production of a p-phenetidine di-imine dimer. The latter could be formed by head to tail coupling of two p-phenetidine radicals which would result in the production of ethanol. A p-phenetidine di-imine dimer could not be isolated from preparative TLC but its formation was confirmed by the use of BHA as a trap-ping agent and subsequent isolation and characterization of the adduct formed (75). Di-imine derivatives react readily with BHA to form adducts (76) due to the electrophilic na-ture of the di-imine nitrogen (77). The structure of the BHA adduct formed (Fig. 6; 75) revealed that the mechanism of the reaction between the di-imine dimer and BHA involved substitution para to the hydroxy group of BHA with the re-sultant elimination of the methoxy substituent of BHA (Fig. 6; 75). This was quite unexpected but has also recently been shown to occur in a reaction of BHA with the di-imine product formed during peroxidase catalyzed oxidation of benzidine (78). The identification of the BHA adduct thus confirmed that a di-imine dimer was formed during peroxida-se catalyzed oxidation of p-phenetidine.

Di-imines are known to be unstable to hydrolysis and de-compose readily in aqueous solution to yield a quinone imi-ne derivative and ammonia. Such a mechanism for the forma-tion of the p-phenetidine quinone imine dimer is supported by the almost stoichiometric production of ammonia and the quinone imine dimer with the concomitant disappearance of the di-imine dimer (Fig. 7; 75). In addition the observa-tion that the combined formation of the quinone-imine dimer and the quinone di-imine dimer correlated very well with the formation of ethanol further strenghtened this hypothe-sis (Fig. 7).

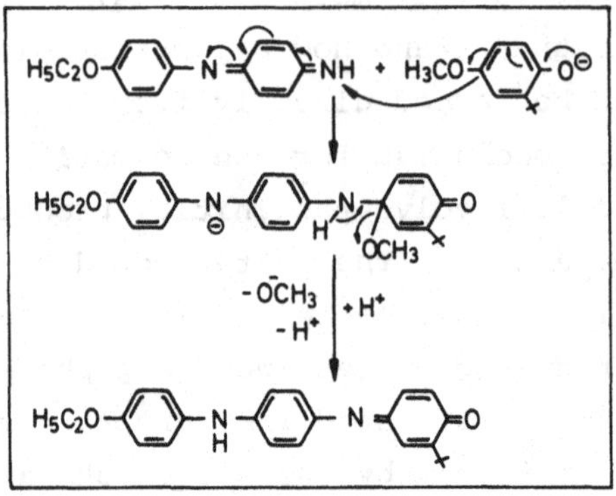

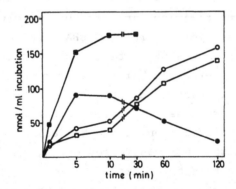

FIGURE 6.   Proposed mechanism for the reaction between p-phenetidine di-imine dimer and BHA.

FIGURE 7.   Time-course of formation of ethanol ▇, di-imine-BHA adduct ●, p-phenetidine-quinone-imine dimer o, and ammonia ▢, during HRP/H_2O_2 catalyzed metabolism of p-phenetidine.

These results suggest the following mechanism for the PGS and HRP catalyzed metabolism of p-phenetidine (Fig. 8). Initially a substrate free radical presumably the amine radical is formed. This radical may be delocalized to form carbon centered radicals. The radical will rapidly oxidize

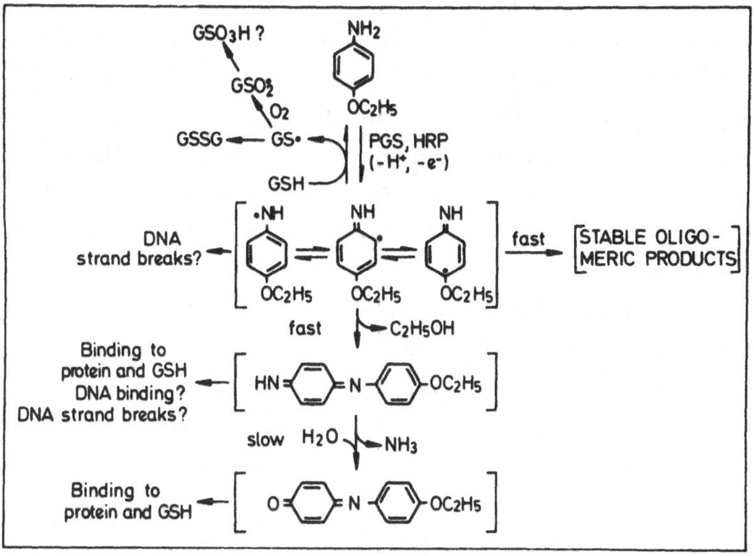

FIGURE 8. Mechanism of the peroxidase catalyzed metabolism of p-phenetidine to reactive products.

GSH with the regeneration of p-phenetidine and formation of a thiyl radical which will either dimerize to GSSG or react with molecular oxygen. The p-phenetidine radical intermediate is very unstable and reacts rapidly through coupling reactions to form several stable and at least one reactive product. Two of these stable oligomeric products have been identified as a diazo- and a trimeric product. Quantitatively the most important product was however the reactive di-imine dimer (75). The di-imine dimer is probably formed through head to tail coupling between two radical intermediates with concomitant loss of ethanol. The p-phenetidine di-imine dimer is then rather slowly hydrolyzed to the quinone imine dimer resulting in elimination of ammonia. This product is stable in the absence of strong nucleophiles and can thus be isolated and characterized.

The relative importance of the radical intermediate, the di-imine dimer and the quinone-imine dimer for the toxic effects observed during the peroxidase catalyzed oxidation of p-phenetidine is not entirely clear. The di-imine and quinone-imine are of similar quinonoid structure and as such would be expected to react with biological nucleophiles in a similar manner. Both of these products are thus probably responsible for the protein and GSH binding observed. DNA binding was not observed with the quinone imine dimer but the di-imine dimer by analogy with the di-imine of benzidine may be able to bind to DNA. In addition the quinone imine does not induce single strand breaks in DNA. Preliminary experiments indicate that a radical intermediate may in fact be responsible for this effect.

The mechanistic studies in peroxidase catalyzed oxidation of p-phenetidine were all performed using either RSVM or HRP. p-Phenetidine is however also metabolized by human kidney medulla microsomes to reactive protein binding product(s) (Table 7; 79). The activity is low but significant and is catalyzed by the peroxidase component of PGS. Furthermore H_2O_2 could also support this PGS dependent protein binding in human kidney.

Table 7. Irreversible binding of (^{14}C) p-phenetidine to human kidney medulla microsomes

Incubation conditions	nmol (^{14}C) p-phenetidine bound/mg protein per 30 min
Complete system	0.50
- Arachidonic acid (AA)	0.23
+ Acetylsalicylic acid, 1 mM	0.22
+ GSH, 1 mM	0.16
- AA + Linolenic acid Hydroperoxide, 100 µM	2.27
- AA + 15-Hydroperoxy-arachidonic acid, 100 µM	0.66
- AA + Hydrogen peroxide, 100 µM	0.70

Incubations contained 0.5 mg microsomal protein, 50 nmoles AA, 50 nmoles (^{14}C) p-phenetidine (1 mCi/mmole), 1 mM EDTA and 100 mM phosphate buffer, pH 8.0 in a total volume of 0.5 ml.

4. CONCLUDING REMARKS

Prostaglandin synthase has the ability to catalyze the metabolism of acetaminophen and the phenacetin metabolite p-phenetidine although phenacetin itself is not a substrate for PGS. A minor metabolite of both acetaminophen and phenacetin, p-aminophenol, is also metabolized by PGS. The metabolism of these three substrates leads to formation of reactive products which have been shown to exert a number of toxic effects.

The extrapolation of in-vitro observations in order to assess potential in-vivo relevance is always very difficult. The situation in-vivo may be completely different and factors such as substrate bioavailability, lipid- and hydrogen peroxide levels, cyclooxygenase and lipoxygenase activity and cellular defence systems are all important factors. Thus on the basis of the data described one cannot propose this type of reaction to be responsible for the nephrotoxic effects of acetaminophen and phenacetin observed in-vivo. There are however some factors which support a possible involvement of PGS in this type of toxicity. The very high affinity of PGS for both acetaminophen and p-phenetidine indicates that even at therapeutic doses this type of metabolic activation may still be active. Furthermore, PGS is located primarily in the renal inner medulla (32) and after the administration of acetaminophen the drug is also localized predominantly in this area of the kidney. In addition the interstitial space has been shown to be the primary target in the development of papillary necrosis in analgesic nephropathy (80). This is of particular interest since the PGS activity is high in the inner medulla interstitial cells (33).

Release and oxygenation of AA might not entirely be the rate limiting factor for the cooxidation of acetaminophen and p-phenetidine. Other lipid hydroperoxides which can also support PGS-dependent cooxidation may be formed through lipoxygenase reactions or via lipid peroxidation. Furthermore hydrogen peroxide can also support PGS-catalyzed metabolism and is formed during several cellular reactions. Involvement

of other hydroperoxides could also explain why acetaminophen and phenacetin induced nephrotoxicity has been observed when they have been taken in combination with the cyclooxygenase inhibitor acetylsalicylic acid.

Phenacetin is known to cause tumours in the kidney and
urinary tract whereas acetaminophen is not. This could possibly be explained by the difference in genotoxicity between the reactive products of acetaminophen and p-phenetidine. Only the products from p-phenetidine induced DNA
single strand breaks in human fibroblasts.

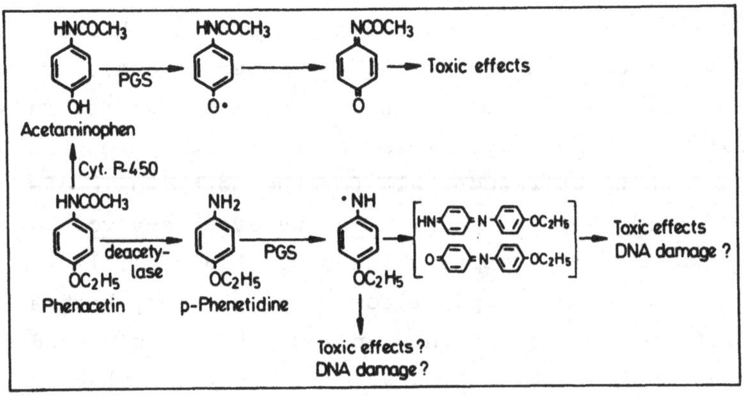

FIGURE 9. Tentative mechanism for acetaminophen and phenacetin induced nephrotoxicity.

The involvement of PGS in the metabolic activation of
acetaminophen and phenacetin can thus be summarized as
shown in Fig. 9. Acetaminophen is metabolized through
an intermediate phenoxy radical, with subsequent formation of NAPQI which is cytotoxic but not genotoxic. Phenacetin is metabolized to acetaminophen which may cause the
nephrotoxic effects observed after phenacetin abuse.
Phenacetin is also metabolized to p-phenetidine, an
intermediate which in turn can undergo further oxidation most likely via a radical intermediate to a di-imine

dimer and a quinone-imine dimer. All of these products seem to have cytotoxic potential but which metabolite or metabolites are responsible for the genotoxic effects observed after PGS-dependent metabolic activation of p-phenetidine is still unclear.

REFERENCES

1. Davidson D, Eastham W: Acute liver necrosis following overdose of paracetamol. Brit Med J (2):497-499, 1966.
2. Brown RAG: Hepatic and renal damage with paracetamol overdosage. J Clin Pathol (21):793, 1968.
3. Boyer TD, Rouff SL: Acetaminophen-induced hepatic necrosis and renal failure. J Amer Med Assoc (218):440-441, 1971.
4. Proudfoot AT, Wright N: Acute paracetamol poisoning. Brit Med J (3):557-558, 1970.
5. Prescott LF, Wright N, Roscoe P, Brown SS: Plasma-paracetamol half-life and hepatic necrosis in patients with paracetamol overdosage. Lancet (1):519-522, 1971.
6. Jollow DJ, Mitchell JR, Potter WZ, Davis DC, Gillette JR, Brodie BB: Acetaminophen-induced hepatic necrosis II, Role of covalent binding in vivo. J Pharmacol Exp. Ther (187):195-202, 1973.
7. Mitchell JR, Jollow DJ, Potter WZ, Davis DC, Gillette JR, Brodie BB: Acetaminophen-induced hepatic necrosis. I. Role of drug metabolism. J Pharmacol Exp Ther (187): 185-194, 1973.
8. Mitchell JR, Jollow DJ, Potter WZ, Gillette JR, Brodie BB: Acetaminophen-induced hepatic necrosis. IV. Protective role of glutathione. J Pharmacol Exp Ther (187): 211-217, 1973.
9. Moldéus P: Paracetamol metabolism and toxicity in isolated hepatocytes from rat and mouse. Biochem Pharmacol (27):2859-2863, 1978.
10. McMurtry RJ, Snodgrass WR, Mitchell JR: Renal necrosis, glutathione depletion and covalent binding after acetaminophen. Toxicol Appl Pharmacol (46):87-100, 1978.
11. Gabriel R, Caldwell J, Hartley RB: Acute tubular necrosis, caused by therapeutic doses of paracetamol? Clin Nephr (18):269-271, 1982.
12. Cobden I, Record CO, Ward MK, Kerr DNS: Paracetamol-induced acute renal failure in the absence of fulminant liver damage. Brit Med J (284):21-22, 1982.
13. Prescott LF, Proudfoot AT, Creegen RJ: Paracetamol-induced acute renal failure in the absence of fulminant liver damage. Brit Med J (284):421-422, 1982.
14. Duggin GG: Mechanisms in the development of analgesic nephropathy. Kidney Int (18):553-561, 1980.
15. Stewart JH: Analgesic abuse and renal failure in Australasia. Kidney Int (13):72-78, 1978.

16. Spühler O, Zollinger HV: Die chronische-interstitille Nephritis. Z Klin Med (151):1-9, 1953.
17. Bengtsson U: A comparative study of chronic non-obstructive pyelonephritis and renal papillary necrosis. Acta Med Scand (Suppl) (388):5-71, 1962.
18. Hultengren N, Lagergren C, Ljungqvist A: Carcinoma of the renal pelvis in renal papillary necrosis. Acta Chir Scand (130):314-320, 1965.
19. Johansson S, Wahlqvist, L: Tumorus of urinary bladder and ureter associated with abuse of phenacetin-containing analgesics. Acta Pathol Microbiol Scand Sect A (85): 768-774, 1977.
20. Isaka H, Yoshii H, Otsuji A, Koike M, Nagai Y, Koura M, Sugiyasu K, Kanabayaishi T: Tumors of Sprague-Dawley rats induced by long-term feeding of phenacetin. Gann (70):29-36, 1979.
21. Johansson SL: Carcinogenicity of analgesics: long-term treatment of Sprague-Dawley rats with phenacetin, phenazone, caffeine and paracetamol (acetaminophen). Int J Cancer (27):521-529, 1981.
22. Smith RL, Timbrell JA: Factors affecting the metabolism of phenacetin. I. Influence of dose, chronic dosage, route of administration and species on the metabolism of $(1-^{14}C\text{-acetyl})$phenacetin. Xenobiotica (4):489-501, 1974.
23. Klutch A, Harfenist M, Conney AH: 2-hydroxy-acetophenetidine, a new metabolite of acetophenetidine. J Med Chem (9):63-66, 1966.
24. Nery R: Some new aspects of the metabolism of phenacetin in the rat. Biochem J (122):317-326, 1971.
25. Calder IC, Creek MJ, Williams PJ, Funder CC, Green CR, Ham KN, Tange JD: N-hydroxylation of p-acetophenetidide as a factor in nephrotoxicity. J Med Chem (16):499-502, 1973.
26. Calder IC, Goss DE, Williams PJ, Funder CC, Green CR, Ham KN, Tange JD: Neoplasia in the rat induced by N-hydroxyphenacetin, a metabolite of phenacetin. Pathology (8):1-6, 1976.
27. Mulder GJ, Hinson JA, Gillette JR: Conversion of the N-O-glucuronide and N-O-sulfate conjugates of N-hydroxyphenacetin to reactive intermediates. Biochem Pharmacol (27):1641-1649, 1978.
28. Vaught JB, McGarvey PB, Lee M-S, Garner CD, Wang CY, Linsmaier-Bednar EM, King CM: Activation of N-hydroxyphenacetin to mutagenic and nucleic acid-binding metabolites by acyltransfer, deacylation and sulfate conjugation. Cancer Res (41):3424-3429, 1981.
29. Wirth PJ, Dybing E, von Bahr C, Thorgeirsson SS: Mechanism of N-hydroxyacetylarylamine mutagenicity in the Salmonella test-system. Metabolic activation of N-hydroxyphenacetin by liver and kidney fractions from rat, mouse, hamster and man. Mol Pharmacol (18):117-127, 1980.
30. Carpenter HM, Mudge GH: Acetaminophen nephrotoxicity: Studies on renal acetylation and deacetylation. J Pharmacol Exp Ther (218):171-167, 1981.

31. Newton JF, Yoshimoto M, Bernstein J, Rush GF, Hook JB: Acetaminophen nephrotoxicity in the rat. II. Strain differences in nephrotoxicity and metabolism of p-aminophenol, a metabolite of acetaminophen. Toxicol Appl Pharmacol (69):307-318, 1983.

32. Zenser TV. Mattammal MB, Davis BB: Differential distribution of mixed function oxidase activities in rabbit kidney. J Pharmacol Exp Ther (207):719-725, 1978.

33. Cavallo T: Fine structural localization of endogenous peroxidase activity in inner medullary interstitial cells of the rat kidney. Lab Invest (31):458-464, 1974.

34. Smith WL, Wilkin GP: Immunochemistry of prostaglandin endoperoxide forming cyclooxygenases: The detection of the cyclooxygenases in rat, rabbit and guinea pig kidneys by immunofluorescence. Prostaglandins (13):873-892, 1977.

35. Samuelsson B, Goldyne M, Granström E, Hamberg M, Hammarström S, Malmsten C: Prostaglandins and thromboxanes. Ann Rev Biochem (47):997-1029, 1978.

36. Miyamoto T, Ogino N, Yamamoto S, Hayaishi O: Purification of prostaglandin endoperoxide synthetase from bovine vesicular gland microsomes. J Biol Chem (251):2629-2636, 1976.

37. O'Brien PJ, Rahimtula A: The possible involvement of a peroxidase in prostaglandin biosynthesis. Biochem Biophys Res Commun (70):832-838, 1976.

38. Vane JR: Inhibition of prostaglandin synthesis as a mechanism of action for aspirin-like drugs. Nature (New Biol) (231):232-235, 1971.

39. Marnett LJ: Polycyclic aromatic hydrocarbon oxidation during prostaglandin biosynthesis. Life Sci (29):531-546, 1981.

40. Marnett LJ, Eling TE: Cooxidation during prostaglandin biosynthesis: A pathway for the metabolic activation of xenobiotics. In: Hodgson E, Bend JR, Philpot RM (ed) Reviews in Biochemical Toxicology 5. Elsevier Biomedical. New York, 1983, pp 135-172.

41. Eling T, Boyd J, Reed G, Mason R, Sivarajah K: Xenobiotic metabolism by prostaglandin endoperoxide synthetase. Drug Metab Rev (14):219-248, 1983.

42. Moldéus P, Rahimtula A: Metabolism of paracetamol to a glutathione conjugate catalyzed by prostaglandin synthetase. Biochem Biophys Res Commun (96):469-475, 1980.

43. Mohandas J, Duggin GG, Horvath JS, Tiller DJ: Metabolic oxidation of acetaminophen (paracetamol) mediated by cytochrome P-450 mixed-function oxidase and prostaglandin endoperoxide synthetase in rabbit kidney. Toxicol Appl Pharmacol (61):252-259, 1981.

44. Boyd JA, Eling TE: Prostaglandin endoperoxide synthetase-dependent cooxidation of acetaminophen to intermediates which covalently bind in vitro to rabbit renal medullary microsomes. J Pharmacol Exp Ther (219):659-664, 1981.

45. Moldéus P, Andersson B, Rahimtula A, Berggren M: Prostaglandin synthetase catalyzed activation of paracetamol. Biochem Pharmacol (31):1363-1368, 1982.

46. Egan RW, Gale PH, Kuehl Jr FA: Reduction of hydroperoxides in the prostaglandin biosynthetic pathway by a microsomal peroxidase. J Biol Chem (254):3295-3302, 1979.
47. Zenser TV, Mattammal MB, Herman CA, Joshi S, Davis BB: Effect of acetaminophen on prostaglandin E_2 and prostaglandin $F_{2\alpha}$ synthesis in the renal inner medulla of rat. Biochim Biophys Acta (542):486-495, 1978.
48. Egan RW, Paxton J, Kuehl Jr FA: Mechanism for irreversible self-deactivation of prostaglandin synthetase. J Biol Chem (251):7329-7335, 1976.
49. Jollow DJ, Thorgeirsson SS, Potter WZ, Hashimoto M, Mitchell JR: Acetaminophen-induced hepatic necrosis. VI. Metabolic disposition of toxic and non-toxic doses of acetaminophen. Pharmacology (12):251-271, 1974.
50. Hinson JA, Pohl LR, Monks TJ, Gillette JR: Acetaminophen-induced hepatotoxicity. Life Sci (29):107-116, 1981.
51. Corcoran GB, Mitchell JR, Vaishnav YN, Horning EC: Evidence that acetaminophen and N-hydroxyacetaminophen form a common arylating intermediate, N-acetyl-p-benzoquinone-imine. Mol Pharmacol (18):536-542, 1980.
52. Nelson SD, Dahlin DC, Rauchman EJ, Rosen GM: Peroxidase-mediated formation of reactive metabolites of acetaminophen. Mol Pharmacol (20):195-199, 1981.
53. Ohnishi T, Yamazaki H, Iyanagi T, Nakamura T, Yamasaki I: One-electron-transfer reactions in biological systems. II. The reaction of free radicals formed in the enzymatic oxidation. Biochim Biophys Acta (172):357-369, 1969.
54. deVries J: Hepatotoxic metabolic activation of paracetamol and its derivatives phenacetin and benorilate: oxygenation or electron transfer? Biochem Pharmacol (30): 399-402, 1981.
55. Duggin GG, Mudge GH: Analgesic nephropathy: Renal distribution of acetaminophen and its conjugates. J Pharmacol Exp Ther (199):1-9, 1976.
56. Andersson B, Nordenskjöld M, Rahimtula A, Moldéus P: Prostaglandin synthetase-catalyzed activation of phenacetin metabolites to genotoxic products. Mol Pharmacol (22):479-485, 1982.
57. Josephy PD, Eling TE, Mason RP: An electron spin resonance study of the activation of benzidine by peroxidases. Mol Pharmacol (23):766-770, 1983.
58. Andersson B, Larsson R, Rahimtula A, Moldéus P: Hydroperoxide-dependent activation of p-phenetidine catalyzed by prostaglandin synthase and other peroxidases. Biochem Pharmacol (32):1045-1050, 1983.
59. Andersson B, Larsson R, Rahimtula A, Moldéus P: Prostaglandin synthase and horseradish peroxidase catalyzed DNA-binding of p-phenetidine. Carcinogenesis (5):161-165, 1984.
60. Baud L, Nivez M-P, Chansel D, Ardaillou R: Stimulation by oxygen radicals of prostaglandin production by rat renal glomeruli. Kidney Int (20):332-339, 1981.

61. Griffin BW, Ting PL: Mechanism of N-demethylation of aminopyrene by hydrogen peroxide catalyzed by horseradish peroxide, metmyoglobin and protohemin. Biochemistry (17):2206-2211, 1978.
62. Marnett LJ, Siedlik PH, Fung LWM: Oxidation of phenidene and BW 755 C by prostaglandin endoperoxide synthetase. J Biol Chem (257):6957-6964, 1982.
63. Josephy PD, Eling T, Mason RP: An electron spin resonance study of the activation of benzidine by peroxidases. Mol Pharmacol (23):766-770, 1983.
64. Josephy PD, Eling T, Mason RP: The horseradish peroxidase catalyzed oxidation of 3,5,3',5'-tetramethylbenzidine:free radical and charge transfer complex intermediates. J Biol Chem (257):3669-3675, 1982.
65. Josephy PD, Eling T, Mason RP: Cooxidation of the clinical reagent 3,5,3',5'-tetramethylbenzidine by prostaglandin synthase. Cancer Res (42):2567-2570, 1982.
66. Ross D, Larsson R, Andersson B, Nilsson U, Lindqvist T, Lindeke B, Moldéus P: The oxidation of p-phenetidine by horse radish peroxidase and prostaglandin synthase and the fate of glutathione during such reactions. Biochem Pharmacol, In press, 1984/1985.
67. Moldéus P, O'Brien PJ, Thor H, Berggren M, Orrenius S: Oxidation of glutathione by free radical intermediates formed during peroxidase-catalyzed N-demethylation reactions. FEBS Lett (162):411-415, 1983.
68. Moldéus P, Jernström B: Interaction of glutathione with reactive intermediates. In: Larsson A, Orrenius S, Holmgren A, Mannervik B (eds). Functions of glutathione. Biochemical, Physiological Toxicological and Clinical Aspects. Raven Press, New York, 1983, pp 99-108.
69. Schafer K, Bonfacic M, Bahnemann D, Asmus KO: Addition of oxygen to organic sulfur radicals. J Phys Chem (82): 2777-2780, 1978.
70. Quintilliani M, Badiello R, Tamba M, Esfandi A, Gorin G: Radiolysis of glutathione in oxygen-containing solutions of pH 7. Int J Radiat Biol (32):195-202, 1977.
71. Quintilliani M, Badiello R, Tamba M, Gorin G: Radiation chemical basis for the role of glutathione in cellular radiation sensitivity. In: Modification of radiosensitivity of biological systems. Vienna IAEA, 1976, pp 29-37.
72. Saunders BC: Peroxidases and catalases. In: Eichhorn GL (ed). Inorganic Biochemistry 2. Elsevier, New York, 1973, pp 988-1021.
73. Parker VD: Anodic oxidation of amines. In: Baizer MM (ed). Organic electrochemistry. Marcel-Dekker Inc., New York, 1973, pp 509-529.
74. Larsson R, Ross D, Rahimtula A, Norbeck K, Lindeke B. Moldéus P: Reactive products formed by peroxidase catalyzed oxidation of p-phenetidine. Chem Biol Interactions, In press, 1984.
75. Ross D, Larsson R, Norbeck K, Ryhage R, Moldéus P: Characterization and mechanism of formation of reactive products formed during peroxidase catalyzed oxidation of p-phenetidine: trapping of reactive species using

reduced glutathione and butylated hydroxyanisole. Molec Pharmacol, In press, 1984.

76. Claiborne A, Fridovich I: Chemical and enzymatic intermediates in the peroxidation of o-dianisidine by horseradish peroxidase. I. Spectral properties of the products of dianisidine oxidation. Biochemistry (18):2324-2329, 1979.

77. Oldfield LF, Bockris JO'M: Reversible oxidation-reduction reactions of aromatic amines. J Phys Colloid Chem (55):1255-1274, 1951.

78. Josephy PD, Van Damme A: Reaction of 4-substituted phenols with benzidine in a peroxidase system. Biochem Pharmacol, submitted, 1983.

79. Larsson R, Andersson B, von Bahr C, Berlin T, Moldéus P: Prostaglandin synthase catalyzed metabolic activation of p-phenetidine in the human kidney medulla. Biochem Pharmacol, submitted, 1983.

80. Saker BM, Kincaid-Smith P: Papillary necrosis in experimental analgesic nephropathy. Brit Med J (1):161-162, 1969.

INDEX